WHAT PEOPLE ARE SAYING ABOUT
YOUR BRAIN AT WORK

"Simply put, this intriguing book offers fascinating research about the brain's functions, limitations, and capacities, and it teaches us how we can 'direct' our own brain chemistry in order to achieve fulfillment and success. Well worth reading and ingesting these skills."
> —Stephen R. Covey, author of *The 7 Habits*
> *of Highly Effective People*

"This is the best, most helpful, and brainiest book I've ever read on how the brain affects how, why, and what we do. After reading only the first four chapters, I felt roughly 100 percent more efficient in organizing my work and personal life. A must-read for anyone who wants to live and work with more disciplined and effective purpose."
> —Warren Bennis, former distinguished professor of business and
> university professor, University of Southern California, and
> author of *On Becoming a Leader*

"This book will improve how you work—by showing you how your brain works!"
> —Marshall Goldsmith, the million-selling author of *Succession:*
> *Are You Ready?* and *What Got You Here Won't Get You There*

"The biggest constraint on most people is the limit of our own brain— our seemingly hardwired neural connections, which make some perceptions and actions feel welcoming and others discomfiting. To accept and then cope with the rote behavior of one's own brain, one must understand such counterintuitive facts as the limits of working memory, the influence of hormones, the power of reappraisal, and much more. *Your Brain at Work* . . . will make a difference to individuals and organizations everywhere."
> —Art Kleiner, former editor in chief of *stategy + business* magazine

"David Rock is just the right guide for bringing the brain to work."
> —Daniel J. Siegel, M.D., clinical professor,
> UCLA School of Medicine

"David has accurately integrated recent neuroscience findings into the business world. These findings provide insights into how to drive change in the brain and improve performance."

—Yi-Yuan Tang, founding director of the Institute of
Neuroinformatics, Dalian University of Technology, China

"If you have no interest in being creative, effective, and happy, then *Your Brain at Work* may not be for you . . . but for everyone else, it's a must-read!"

—Chris Wink, cofounder, the Blue Man Group

"Leadership belongs to the most aware. *Your Brain at Work* gives you the practical and profound tools to build awareness for yourself, your relationships, and your world. David Rock's incredible work allows us to lead with our whole brains—inside and out!"

—Kevin Cashman, Global Leader of CEO & Executive
Development at Korn Ferry, founder of LeaderSource,
and author of *Leadership from the Inside Out*

Your
Brain at
Work

Revised and Updated

Also by Dr. David Rock

Quiet Leadership: Six Steps to Transforming Performance at Work

Personal Best:
Step by Step Coaching for Creating the Life You Want

Coaching with the Brain in Mind: Foundations for Practice

Your Brain at Work

Revised and Updated

STRATEGIES FOR OVERCOMING
DISTRACTION, REGAINING FOCUS, AND
WORKING SMARTER ALL DAY LONG

Dr. David Rock

HARPER
BUSINESS

An Imprint of HarperCollins*Publishers*

*To Lisa, Trinity, and
India Rock*

HarperCollins books may be purchased for educational, business, or
sales promotional use. For information, please email the Special Markets
Department at SPsales@harpercollins.com.

SCARF® is a registered trademark of the NeuroLeadership Institute internationally.

ORIGINALLY PUBLISHED IN 2009 BY HARPER BUSINESS, AN IMPRINT OF
HARPERCOLLINS PUBLISHERS.

Revised and Updated Edition published 2020.

Designed by Nicola Ferguson

Library of Congress Cataloging-in-Publication Data

Names: Rock, David, author.
Title: Your brain at work : strategies for overcoming distraction,
    regaining focus, and working smarter all day long / Dr. David Rock.
Description: Revised and updated [edition]. | New York, NY : Harper Business,
    2020. | Includes bibliographical references and index.
Identifiers: LCCN 2020008535 (print) | LCCN 2020008536 (ebook) | ISBN
    9780063003156 (hardcover) | ISBN 9780063003163 (ebook)
Subjects: LCSH: Success in business--Psychological aspects. | Brain. |
    Distraction (Psychology)
Classification: LCC HF5386 .R538 2020 (print) | LCC HF5386 (ebook) | DDC
    658.4/092019—dc23
LC record available at https://lccn.loc.gov/2020008535
LC ebook record available at https://lccn.loc.gov/2020008536

20 21 22 23 24   LSC   10 9 8 7 6 5 4 3 2 1

CONTENTS

ACT IV: FACILITATE CHANGE

FOREWORD

When I first read the manuscript for *Your Brain at Work*, I asked David Rock if I could share it with my wife and two teenagers. The writing is clear, the ideas transformative, and the setup fabulous: Scenes are played out in normal day-to-day work and home life, then the same situation is reenacted when the characters have learned to think and behave with the "brain in mind." When they've developed the ability to understand their minds more deeply—to have what I call "mindsight"—they have a conscious choice over how to engage their brains, and now have the power to change their habits.

The mind—how we regulate the flow of energy and information—uses the brain to create itself. For this reason, the emerging science of the brain is a natural place to develop more effective strategies for improving life at work. David Rock has taken challenging areas of neuroscience and cognition and interpreted them in an accurate yet highly accessible manner. He has interviewed scientists directly, visiting their laboratories, and has spent hundreds of hours culling the latest findings to extract the most current understanding of how the mind and brain influence our lives.

The suggestions here are powerful tools, based on rigorous science, for helping people throughout the workforce. If you are a frontline employee, the stories and science in this book will help you become more effective in your work and avoid burnout. If you are a manager, the information offered here will enable you to delegate with more skill and juggle various projects with more success. And if you are in a leadership position, knowing about the brain can help you create an organizational structure that inspires your employees to take pride in their work, bring more attention and resourcefulness to their tasks, and work more collaboratively with their peers.

Learning to live with the brain in mind is a powerful way to strengthen your mind and improve your professional life. With more ability to regulate the flow of energy and information in your work, you can become more effective and achieve a greater sense of satisfaction. David Rock is just the right guide for bringing the brain to work. We can all thank him for sharing his hard-earned insights—and wonderful sense of humor.

—*Daniel J. Siegel, M.D., clinical professor at UCLA School of Medicine; co-director of the UCLA Mindful Awareness Research Center; director, Mindsight Institute; and author of* Mindsight: The New Science of Personal Transformation; The Mindful Brain: Reflection and Attunement in the Cultivation of Well-Being; *and* The Developing Mind

INTRODUCTION

An avalanche of emails.
An impossible number of text messages.
Dozens of alerts from Facebook, LinkedIn, and your CRM.
A meeting schedule that leaves you exhausted by 11:00 a.m.
Ever more change and uncertainty as your job changes monthly.
The occasional win just to keep you going.

If this sounds like your average day at work, you've picked up the right book.

This book will help you be more focused and productive, work smarter, stay cool under pressure, reduce the length of meetings, and even tackle the hardest challenge of all: influencing other people. Along the way it may help you be a better parent and partner, and perhaps even live longer. It will even make coffee for you. Okay, maybe not that last bit, but everything else I mean quite seriously.

This book will transform your performance at work by letting you in on recent and important discoveries about the human brain. You'll have the chance to get more focused and productive by understanding your own brain at work—at work. It's only through knowing your brain that you can change it. (How your brain can change by understanding itself is something you will learn about here, too.)

I know a lot about how brains get easily overwhelmed, so I have no intention of drowning you in complicated science here. Instead, in this book you'll get to know your brain in a way that brains like: by reading a story. This story involves two characters, Emily and Paul, as they experience a set of challenges over a single day at work. As you watch Emily and Paul go through their day, some of the smartest neuroscientists in the world will explain why they struggle with their email, schedules, and colleagues. Even better, you will also get to see what Emily and Paul might have done differently if they'd understood their brains better.

Before I explain how this book is organized, let me give you a little background as to how this all came about. I help organizations such as BlackRock, IBM, and Microsoft improve their performance. Over the course of a decade of this work, I've discovered, somewhat by accident, that teaching people managers and employees about their brains made a big difference to their performance, and often to their lives, too. When I couldn't find a book that explained the most useful discoveries about the brain in simple language for people at work, I decided to write one.

This book took three years to put together initially, though I had been developing parts of it for several years longer. It is based on interviews with thirty leading neuroscientists from the United States, Europe, and Asia-Pacific, and it draws from more than three hundred research papers based on thousands of brain and psychological studies done over recent years. While writing this book, I had a scientific mentor help me wade through the research, neuroscientist Dr. Jeffrey M. Schwartz. I also convened three summits about the brain in the workplace: in Italy, Australia, and the United States. Out of those summits, I helped create an academic journal and gave hundreds of lectures and workshops across the globe. Since the first edition of this book came out, I have now run fourteen international summits and written or edited over fifty papers for academic journals, helping to define a whole field of study called NeuroLeadership. The ideas in this book emerged out of a combination of all these activities.

Enough about me. Let's explore how the book is organized. I wanted this book to be useful to people. That's tricky when you're dealing with the most complex thing in the known universe, the human brain. After several attempts to explain the brain in different ways, I decided to structure this book like a play.

The play has four acts. The first two acts are about your own brain. The second two acts focus on interacting with other people's brains. There is also an intermission, which explores some of the deeper themes emerging out of the story.

Act 1 is called "Problems and Decisions," and involves the fundamentals of thinking. Act 2, "Stay Cool Under Pressure," explores emotions and motivations and the impact of these on thinking. Act 3, "Collaborate with Others," introduces research on how we might all get along better. Act 4, "Facilitate Change," focuses on how to create change in others, one of the hardest things to do.

Each act has several scenes, and each scene begins with Emily or Paul facing a challenge at work or at home, such as dealing with an overwhelming volume of information to process first thing in the morning. I chose the characters' particular daily challenges by gathering information, with an online survey I created. I then combined the resulting data with research from surveys of organizational culture.

After watching Emily or Paul go through a challenge in each scene, you get to find out what is going on inside her or his brain that is making life so difficult, and to hear directly from the neuroscientists I interviewed and from other relevant studies. The most fun part of the book is the "take two" at the end of each scene. In each take two, Emily and Paul understand their brains more and, as a result, make a different set of decisions moment to moment. The differences between take one and take two come from tiny changes in behavior, but these changes generate substantially different outcomes. Subtle internal changes, which happen within a fraction of a second and may not be noticeable to the outside world, can sometimes change everything. This book will help you understand, isolate, and reproduce such changes.

At the end of each scene, I summarize the big surprises emerging from brain research. If you want to use this book to change your brain more deeply, each scene includes a list of specific things to try for yourself.

The book finishes with an "encore," which summarizes the science and looks at the bigger implications of the research. I also include a list of further resources, and an extensive annotated bibliography about the studies I have drawn from. I make clear where and how I drew my own conclusions; otherwise, the ideas here are drawn from hundreds of scientific studies that you can read, too, if you like.

The performance starts soon, so it might help to know a little about the main characters and the setting. Emily and Paul are in their early forties. They live in a midsize city with their two teenage children, Michelle and Josh. Emily is an executive in a company that runs large conferences. Paul works for himself as an IT consultant, having spent his earlier years in a larger firm.

The action happens across a single day, an average Monday, normal except that it's Emily's second week of a new promotion. She now has a bigger budget and a larger team to manage. She is excited about her new role and wants to kick things off well, but she has some new skills to learn. Paul is pitching a new project that he hopes will help him grow out of the small home office he has had for five years. The two have plenty of other hopes and dreams, including bringing up their children well, despite their hectic work schedules.

Let's raise the curtain and begin the show.

ACT I

• • • • • • • •

Problems and Decisions

More people than ever are being paid to think, instead of just doing routine tasks. Yet making complex decisions and solving new problems is difficult for any stretch of time because of some real biological limits on your brain. Surprisingly, one of the best ways to improve mental performance is to understand these limits.

In act 1, Emily discovers why thinking requires so much energy, and develops new techniques for dealing with having too much to do. Paul learns about the space limits of his brain, and works out how to deal with information overload. Emily finds out why it's so hard to do two things at once, and rethinks how she organizes her work. Paul discovers why he is so easily distracted, and works on how to stay more focused. Then he finds out how to stay in his brain's "sweet spot." In the last scene, Emily discovers that her problem-solving techniques need improving, and learns how to have breakthroughs when she needs them most.

SCENE 1

• • • • • • • •

The Morning Information Overwhelm

t's 7:30, Monday morning. Emily gets up from the breakfast table, kisses Paul and her children goodbye, closes the front door, and heads to her car. After sorting out sibling squabbles all weekend, she is looking forward to focusing on her new job. As she heads toward the freeway, she thinks about her week ahead and how she wants to get off to a good start. About halfway to work, she gets an idea for a new conference, and has to concentrate hard to keep the idea in mind as she drives.

Emily is at her desk by eight o'clock. She turns on her computer, ready to flesh out this new conference idea. Then, she spies the awful truth: two hundred emails start to download. She has over fifty messages from her company chat, plus dozens of alerts from two other programs. A wave of anxiety washes over her. Answering the emails alone could take all day, but she also has hours of meetings booked and three projects due by six o'clock. Her excitement about the promotion is already beginning to fade. She loves the idea of the extra money and responsibility, but she isn't sure how she's going to cope with the increased workload.

Thirty minutes later, Emily realizes she has responded to only twenty emails, and hasn't touched the chats people sent her, which could be urgent. She needs to speed up. She tries to read emails while listening to her voice mail. Her attention shifts for a moment to how her longer hours might impact her kids. She remembers how she snapped at them in the past when she was too busy at work. Then she remembers a promise she made to herself—to be a good role model by staying true to her career ambitions. Lost in thought, she accidentally deletes a voice mail from her boss.

The burst of adrenaline triggered by the lost message snaps her focus to the present. She stops typing and tries to think about the projects due today: writing a new conference proposal, crafting some marketing copy, and deciding which assistant to hire. Then there are all those emails, with dozens of different issues to follow up on. She spends several seconds trying to imagine how to prioritize everything, but nothing comes to mind. She tries to remember the guidelines she learned in a time-management course she took long ago, but despite a few seconds of focus, she can't find the thread of the memory. She goes back to the emails and tries to type faster.

By the end of the hour, Emily has replied to forty emails and juggled the critical instant messages, but with the workday starting, there are now 120 emails waiting. She's had no time to work on her new conference idea, either. Despite her good intentions, it's not a great start to the day, the week, or her new position.

Emily is not alone. Workers everywhere are experiencing an epidemic of overwhelm. For some people, it's the pressure of a promotion; for many others, a downsizing or reorganization; but for many, every day involves a constant, massive, and overwhelming volume of work. As the world digitizes, globalizes, unplugs, and reorganizes, having too much to do has become our biggest complaint.

For Emily to be effective in her new job, without destroying her health or her family, she needs to change how her brain works. She needs new neural circuitry for managing a significantly larger and more complex to-do list.

The trouble is, when it comes to making decisions and solving problems, as Emily is trying to do this morning, the brain has some surprising performance limitations. While the brain is exquisitely powerful, even the brain of a Harvard graduate can be turned into that of an eight-year-old simply by being made to do two things at once. In this scene, and in the next few to come, Emily and Paul are going to discover the biological limits that underlie mental performance, and in the process develop more brain-smart approaches to everyday challenges. As they change their brains, you will have the opportunity to change yours, too.

THE GOLDILOCKS INSIDE US ALL

Making decisions and solving problems relies heavily on a region of the brain called the prefrontal cortex. The cortex is the outer covering of the brain, the curly gray stuff you see in pictures of brains. It's a tenth of an inch thick and covers the brain like a sheet. The prefrontal cortex, which sits behind the forehead, is just one part of the overall cortex.

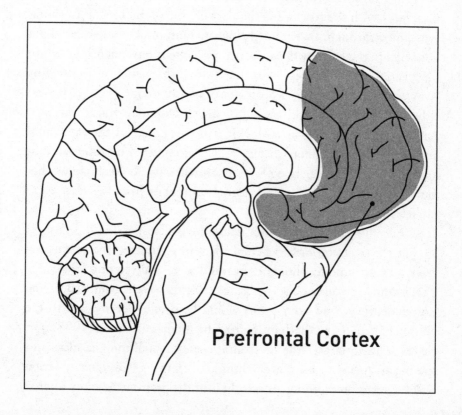

Prefrontal Cortex

The last major brain region to develop during human evolutionary history, it is a measly 4 to 5 percent of the volume of the rest of the brain.

Don't be fooled, though. As with diamonds and espressos, sometimes good things come in small packages. Without a prefrontal cortex you wouldn't be able to set any type of goal. Thinking "Get some milk at the store" would be impossible. You also wouldn't be able to plan. You wouldn't be able to say to yourself, "Walk up the hill, go into the store, and purchase the milk, then walk back down." You wouldn't be able to control impulses, so if you felt an urge to lie on a sun-warmed road on a cold day, you'd be in trouble. And you wouldn't be able to solve problems, such as working out how to get to a hospital after a car has run you over. You would also have trouble visualizing a situation you'd never seen before, so you'd have no idea what to take to the hospital. And, finally, you wouldn't be able to think creatively, so you wouldn't be able to make up a good story to tell your wife when you finally got home from the hospital.

Your prefrontal cortex is the biological seat of your conscious interactions with the world. It's the part of your brain central to thinking things through, instead of being on "autopilot" as you go about your life. In the last few decades, neuroscientists have made important discoveries about this region of the brain, in particular a team led by Amy Arnsten. Arnsten is a professor of neurobiology at Yale Medical School. Like her mentor, the late Patricia Goldman-Rakic, Arnsten has devoted her career to unlocking the mysteries of the prefrontal cortex. "Your prefrontal cortex holds the contents of your mind at any one point," Arnsten explains. "It's where we hold thoughts that are not being generated from external sources or from the senses. We ourselves are generating them."

While the prefrontal cortex is handy, it also has big limitations. To put these limitations in perspective, imagine that the processing resources for holding thoughts in mind were equivalent to the value of the coins in your pocket right now. If this were so, the processing power of the rest of your brain would be roughly equivalent to the entire U.S. economy (perhaps before the financial crisis of 2008). Or, as Arnsten explains, "the prefrontal cortex is like the Goldilocks of the brain. It has to have everything just right or it doesn't function well." Getting everything "just right" for the prefrontal cortex is what

Emily needs to learn to do, to get on top of the extra information she is juggling in her new job.

THE STAGE

I am going to introduce a metaphor for the prefrontal cortex that will be used throughout the book. Think of the prefrontal cortex as a stage in a small theater, where actors all play a specific part. The actors in this case represent information that you hold in your attention. Sometimes these actors enter the stage as a normal actor would, from the side of the stage. This is the case when information from the outside world comes to your attention, such as when Emily watches her computer download hundreds of emails.

However, this stage is not exactly like the stage in an ordinary theater. Sometimes the actors might also be audience members who get onstage to perform. The audience represents information from your inner world: your own thoughts, memories, and imaginings. The stage is what you focus on at any one time, and it can hold information from the outside world, information from your inner world, or any combination of the two.

Once actors get on the stage of your attention there are lots of interesting things you can do with them. To *understand* a new idea, you put new actors on the stage and hold them there long enough to see how they connect to audience members—that is, to information already in your brain. Emily does this when she reads each email to understand its contents, and hopefully you are doing this right now with this book. To make a *decision*, you hold actors onstage and compare them to one another, making value judgments. Emily does this as she reads each email and decides how to respond.

To *recall* information, meaning to bring a memory from the past back to mind, you bring an audience member up on the stage. If that memory is old, it might be at the back of the audience, in the dark. It can take time and effort to find this audience member, and you might get distracted along the way. Emily struggles with trying to remember rules for managing emails from a training course, but the information is too far back in the audience, so she gives up. To *memorize* information, you need to get actors off the stage and into the audience. Emily tries to memorize an idea for a new conference while driving, but finds this effort tiring.

Sometimes it's important *not* to focus on an actor, to keep him off the stage. For example, you might have a tight deadline at lunchtime and are trying to focus on a project, but you find the thought of lunch keeps jumping into your awareness, distracting you for half a minute each time. The process of *inhibition*, of keeping certain actors off the stage, requires a lot of effort. It's also central to effective functioning in life. Emily, distracted as she ruminates on how she is going to cope in her new job, accidentally deletes a voice mail as a result.

These five functions, *understanding, deciding, recalling, memorizing,* and *inhibiting,* make up the majority of conscious thought and are recombined to plan, problem-solve, communicate, and other tasks. Each function uses the prefrontal cortex intensively and requires significant resources to operate, far more resources than Emily realizes.

THE STAGE NEEDS A LOT OF LIGHTING

Recently my wife and I walked up the hill to the local shops—for some milk, funnily enough—and my wife asked me a question that I had to stop walking to answer. Everyone knows that walking up a hill takes energy. It turns out that conscious mental activities do, too, and I didn't have enough energy to do both.

Conscious mental activities chew up metabolic resources, the fuel in your blood, significantly faster than automatic brain functions such as how your brain helps keep your heart beating or your lungs breathing. The stage requires a lot of energy to function. It's as if the lights are a long way back from the stage, so you need a lot of them, all on full, to see the actors. To make matters worse, the power to light the stage is a limited resource, decreasing as you use it, a bit like a set of batteries that constantly need recharging.

The first clinical evidence for this limitation came way back in 1898. The scientist J. C. Welsh measured people's ability to do physical tasks while thinking. She had subjects start a mental task and then asked them to impose as much force as they could, at the same time, on a dynamometer, a machine for measuring force. Her measurements showed that almost all mental tasks reduced maximum force, often by as much as 50 percent.

Doing energy-hungry tasks with your stage, such as scheduling meetings, might exhaust you after just an hour. In comparison, a truck driver can drive all day and night, his or her other ability to keep going limited only by his or her need for sleep. Driving a truck doesn't require much use of the prefrontal cortex (unless you are a new driver, in a new truck, or on a new route). It involves another part of the brain called the basal ganglia. The basal ganglia are four masses in the brain driving routine activities that don't require a lot of active mental attention. From an evolutionary perspective, the basal ganglia are an older part of the brain. They are also highly energy efficient, with fewer overall limitations than the prefrontal cortex. As soon as you repeat an activity even just a few times, the basal ganglia start to take over. The basal ganglia, and many other brain regions, function beneath conscious awareness, which explains why Emily can drive and think about a conference at the same time.

The prefrontal cortex chews up metabolic fuel, such as glucose and oxygen, faster than people realize. "We have a limited bucket of resources for activities like decision making and impulse control," Dr. Roy Baumeister from Florida University explains, "and when we use these up, we don't have as much for the next activity." Make one difficult decision, and the next is more difficult. This effect can be fixed by drinking a glucose drink. Baumeister tested this hypothesis using lemonade sweetened with either glucose or a sugar substitute, and the impact on performance was marked.

Baumeister's insight is a significant discovery about the machinery of the brain. Your ability to operate the stage has real limits because the stage needs a lot of fuel. It requires a lot of power to run, and this power drains as you use it. This explains many everyday phenomena such as why it's easy to get distracted when you're tired or hungry. When you get to two o'clock in the morning and can't seem to think, it's not you—it's your brain. Your best-quality thinking lasts for a limited time. The answer is not always just to "try harder."

Why does the mental stage require so much energy to function?

Some scientists think the prefrontal cortex is energy hungry because it is still new in evolutionary terms and needs to evolve further to meet today's information demands. Here's a different perspective. When you understand the brain processes involved in an activity such as decision-making, you might be amazed at the capacity you do have. You might respect your limitations rather than fight them. Let's explore this idea by going back briefly to another step in Emily's story.

Emily walks into a conference room at 9:00 a.m. Her brain takes in a flood of data: a cacophony of sounds as three people speak at once, a vibrant mash of colors from flipcharts, people's suits, and art on the walls, a multitude of shapes and movement, a dozen faces. The volume and complexity of information entering her brain at that moment would be enough to freeze any supercomputer. As Emily walks into the room, she uses her *short-term memory* to process incoming information. Large volumes of data enter her brain, but twenty to thirty seconds later, most of this data has gone. It's like hundreds of new actors jumping on to the stage briefly and then running off. If you asked Emily a minute later what she had seen, she couldn't tell who was wearing what, or describe what was on a flipchart, unless she stopped to pay close attention and noted these things in particular.

A moment later, Emily remembers why she has come there in the first place—to meet a new colleague named Madelyn for coffee. Her brain now has to manage three energy-hungry processes at once. These three processes involve many parts of her brain, but her prefrontal cortex manages the overall process. First, incoming data about the room, both visual and auditory, continue flowing into her short-term memory, yet now this data must be observed, in the way you might look at a car in the parking lot to see if it's yours. You have to hold the data on the stage, which requires effort. This effort consumes energy.

Second, Emily must bring an image of Madelyn onto her stage, to compare the incoming information about the room to something. The image for Madelyn's face is dredged out of trillions of bits of data held in Emily's long-term memory. Emily needs to keep the circuits representing a picture of Madelyn active, to keep this actor on the stage, which also requires effort, and more energy.

Finally, Emily must keep in mind the idea of "coffee." Otherwise, when she finds Madelyn, she will forget why she was looking for her.

These three processes—"incoming data closely observed," "Madelyn," and "coffee"—all need to remain active at the same time. Meanwhile, new data continues to enter her short-term memory that might disrupt these processes. There are now three groups of actors that require energy for Emily to keep them onstage, with new actors scrambling to get onstage that need to be kept off.

Emily finds Madelyn. "Where should we go?" Madelyn asks as they leave the room. "No idea. I can't think of anything just yet," Emily replies. "Let's walk around and see if we can find somewhere to sit and think."

What does Emily's story mean? Perhaps you are now "seeing" (holding something on your own stage) that the mental stage is a hungry animal. You can take this information several ways. One option involves bemoaning the state of human functioning. The other option could be to send your assistant out for some glucose powder or today's ready-made solution: a cola drink. (This option may help, though with some unfortunate possible side effects such as gaining weight, adding to your dental costs, or increasing the risk of Type 2 diabetes.) A third option, and the one I would suggest, is to rethink how you value and use this resource called the mental stage.

If the mental stage is a limited resource, it's like other limited resources such as stocks, gold, or cash. Imagine if Emily treated her capacity to think in the same way her company managed its financial assets, with a tight control on spending. Instead, Emily wastes her resources by trying to hold an idea in mind about a new conference while she's driving, tiring her brain before she even gets to work. Then she begins the morning by working through her emails. Processing large amounts of information uses a lot of resources, probably not the best use of her most important asset right then.

Here's a new perspective: each time you use your mental stage, allocate it to something important. It is a limited resource that can't be wasted. No matter how much effort you put in, you can't sit there and make brilliant decisions all day the way a truck driver can stay on the road.

PRIORITIZE PRIORITIZING

If Emily knew how energy-hungry her stage was, she would start her Monday morning differently. The big difference is she would prioritize prioritizing. She would prioritize first, before any other attention-rich activity such as emailing. That's because prioritizing is one of the brain's most energy-hungry processes.

After even just a few mental activities, you may not have the resources left to prioritize. Using your stage for something energy intensive such as prioritizing is like flying one of those toy helicopters you see in parks, the ones that are supposed to be for kids but that dads actually buy for themselves. Once Dad gets the helicopter off the ground a few times, it won't get off the ground again because the power is too low. It gets close, rising a few inches off, and then collapses back down. And the more you try, the less energy there is. Best to recharge and try again later. In a similar way, doing ten minutes of emailing can use up the power needed for prioritizing. Emily experienced this when she couldn't "see" how to prioritize her day and ended up dealing with her emails instead. To understand why prioritizing is such a hungry beast, let's explore a new idea: the varying degrees of difficulty of getting actors on the stage.

SOME ACTORS ARE HARDER THAN OTHERS TO GET ONSTAGE

This is a useful insight about the brain with broad implications, so bear with me here. It's easy to bring to mind something that's just happened. The circuit is easy to access, as it's "fresh," such as finding an audience member in the front row. Let's try an experiment to make this real. Try to see, in your mind's eye, what you ate for your last meal. Usually this takes just a moment, and little effort. Bringing recent events onto the stage is a relatively fast, low-energy activity. It's like bringing an audience member from the front row onto the stage.

Now picture what you ate for lunch ten days ago. Unless you have a consistent pattern to draw on ("I always have tuna fish sandwiches"), picturing that meal takes several moments longer and a lot more effort than recalling a recent meal. The circuits involved in picturing the earlier lunch are farther back in the audience, so you have to take more time to scan the crowd to find them. Memory researchers show that recalling earlier memories requires tracing back in time, recalling in chronological order the events between now and when the memory was first formed. The farther back a memory is, such as Emily's tips from a training course on dealing with email, the longer this task will take, and the more attention and energy it will require.

Now picture yourself preparing lunch for six in a Japanese restaurant in China. Easy, if you're a Japanese chef who has worked in China! For the rest of us, without any ready-made images in the audience, we have to find suitable audience members and put them together to represent the image of the lunch. You might find a visual for a restaurant, and then find images of six friends, then picture an image from China. It's like trying to bring twenty characters to the stage instead of one. It takes a lot more time and effort. The brain likes to minimize energy usage because the brain developed at a time when metabolic resources were scarce. So there is a slight discomfort involved in putting effort into thinking, or any other activity that uses metabolic resources. (If effort were fun, most households wouldn't have a remote control for the TV, power windows on the car, or dishwashing machines.)

Picturing something you have not yet seen is going to take a lot of energy and effort. This partly explains why people spend more time thinking about problems (things they have seen) than solutions (things they have never seen). It explains why setting goals feels so hard (it's hard to envision the future). Daniel Gilbert's 2006 book, *Stumbling on Happiness*, dives deeply into the implications of this finding, illustrating how human beings are terrible at estimating emotions in the future, a concept he calls *affective forecasting*. Gilbert shows how people define how they will feel in the future based more on the way they feel today, instead of correctly assessing the mental state they might be in at a future date. That's because it's difficult.

This of course also explains why prioritizing is so hard. Prioritizing involves imagining and then moving around concepts of which you have no direct experience. How can Emily decide whether hiring a new assistant is going to be easier than writing a proposal for a conference?

She hasn't seen either event in actuality, so neither event is in her audience. What's more, prioritizing involves every function I mentioned earlier: understanding new ideas, as well as making decisions, remembering, and inhibiting, all at once. It's like the triathlon of mental tasks.

Like the coding used on the ski slopes for degree of difficulty, there are mental tasks that are green, blue, and black. Prioritizing, at least in a knowledge economy full of conceptual projects, is definitely a black run, perhaps even a double black diamond. Do it when you are fresh and energized, or you might crash and burn down the hill.

USE VISUALS

Clearly it's important to prioritize prioritizing. Now, assuming Emily did tackle prioritizing first, with a fresh mind and lots of glucose to access, what else might she do to maximize her ability to prioritize? One way to reduce the energy required for processing information is to use visuals, to literally see something in your mind's eye. For example, right now you are learning about a complex scientific idea, the functioning of the prefrontal cortex, using the metaphor of a stage. Picturing a concept activates the visual cortex in the occipital lobe, at the back of the brain. This region can be activated through actual pictures, or through metaphors, and storytelling, anything that generates an image in mind.

There are a couple of reasons why visuals are so useful. First, they are highly information-efficient constructs. If you picture your bedroom, when you hold the image in mind, that image contains a huge amount of information involving complex relationships among dozens of objects, their sizes and shapes, their relative positions, and so on. Putting all that information into words would take significantly more energy than visualizing it.

Another reason visuals are so helpful is that the brain has a long history of creating mental imagery involving objects and people interacting. Visual processes evolved over millions of years, so the machinery is highly efficient, especially in comparison to the circuitry involved in language. Studies have shown that when you give people a logic problem to solve, they do so dramatically faster when the problem is explained in terms of people interacting rather than in terms of disembodied conceptual ideas.

GET THINGS OUT OF YOUR HEAD

Creating visuals for complex ideas is one way to maximize limited energy resources. Another way involves reducing the load on the prefrontal cortex whenever possible. If Emily gets a piece of paper and writes down the four big projects for the day, she saves her brain for comparing the elements instead of using energy to hold each one. The same benefits can be achieved by using physical objects, such as a stapler, pen, and ruler to represent each project. The idea is to get the concepts out of your mind and into the world, and to save the stage for the most important functions. Minimize energy usage to maximize performance.

If Emily had prioritized first thing in the morning, and got things out of her mind and into the world to compare them, there's still one other thing she could have done here to be most effective this morning. The stage uses up power quickly, and as the lights dim, it gets harder to hold actors in the right place and stop others from getting on the stage. This tendency means scheduling the most attention-rich tasks when you have a fresh and alert mind. This could be early in the morning, or perhaps after a break or exercise. The prefrontal cortex has much in common with other energy-hungry body parts such as muscles. It tires from use, and can do a lot more after a good rest. Making a tough decision might take thirty seconds when you are fresh and be impossible when you're not.

It's helpful to become aware of your own mental energy needs and schedule accordingly. Experiment with different timings. One technique is to break work up into blocks of time based on type of brain use, rather than topic. For example, if you have to do some creative writing in several different projects, which requires a clear, fresh mind, you might do all your creative writing on a Monday. People don't tend to do this—they tend either to work on one project at a time, or to respond to issues as they arise, sometimes thinking at a high abstract level, sometimes at a more detailed level, and then sometimes multi-tasking and switching around a lot. Instead, you could divide a day into blocks of time when you do deep thinking such as creative writing, other blocks for having meetings, and other blocks for routine tasks such as responding to emails. Deep thinking tends to require more effort, so plan to do your deep thinking in one block, perhaps early in the morning or late at night. One big advantage of this strategy is that you can shift around the type of work you do, to let your brain recover. If you were doing physical exercise, you wouldn't do heavy lifting all day. You'd do some heavy lifting, then some cardiovascular exercise, and then some stretching. Each time you changed your exercise mode, your muscles would get used in new ways, with some resting while others worked. It's similar with mixing up types of thinking. Give your brain a rest when you can by mixing things up.

One final insight about prioritizing involves getting disciplined about what you *don't* put on the stage. This means *not* thinking when you don't have to, becoming disciplined about not paying attention to non-urgent tasks unless, or until, it's truly essential that you do. Learning to say no to mental tasks that are not among your priorities is difficult but very helpful. Another technique for thinking less about unnecessary tasks is to delegate well. How do you know what to delegate and what not to delegate? This task, like prioritizing, uses a lot of energy, so is also best done with a fresh mind. Another technique is not to think at all about a project until all the information is at hand. Don't waste energy solving a problem you know you will have more information about later. The bottom line to all this is one simple message: your ability to make great decisions is a limited resource. Conserve this resource at every opportunity.

Now, let's bring the ideas from this chapter together and explore how Emily might have done things differently if she had understood the limitations of her prefrontal cortex.

THE MORNING INFORMATION OVERWHELM, TAKE TWO

It's 7:30, Monday morning. Emily gets up from the breakfast table, kisses Paul and the kids goodbye, and heads toward her car. After a weekend of sibling squabbles, she is looking forward to focusing on her new job. As she heads toward the freeway, she thinks about how she can put on her best performance this week. She gets an exciting idea about a new conference and quickly records these thoughts on her iPhone, using her voice to activate the program while at a traffic light. She knows she shouldn't tire her brain trying to remember things. She puts on the radio and enjoys some good music, letting herself relax.

Emily is at her desk by eight o'clock. She turns on her computer, ready to work on the new conference. Then a few hundred emails download and she sees a pile of alerts and instant messages, and a wave of anxiety washes over her. The stress of the increased workload starts to displace any excitement about her promotion. She loves the idea of the extra money and responsibility, but isn't exactly sure how she's going to cope. The emails alone could take all day, but she has hours of meetings booked and three projects due by six o'clock.

With her anxiety levels rising, Emily decides that prioritizing is essential, but she knows this will take a lot of effort. She closes her computer, turns off the phone, and goes to the whiteboard. Though she is curious to know what's in her emails, she knows there shouldn't be anything that can't wait till later. She consciously stops focusing on the waiting emails. On the whiteboard she creates three small boxes for each of her projects: "conference," "hire assistant," "writing," and another box with the words "email catchup." Then she remembers her new conference idea and writes this up, too.

Emily saves her energy for comparing, rather than holding the concepts on her stage. This small thing makes a big difference: all her processing power is available for considering the relationship between the items. She looks at each box and steps back to find any patterns. The hardest project, she realizes, will be to hire the assistant. She decides to focus on this first. She spends the next forty minutes completing a review of the job applicants so that by the end of the day she can make her decision. She decides to spend her last ten minutes checking emails to see if anything is urgent, just in case.

By the end of the hour, Emily has worked out her preferred candidate for the assistant job and has arranged for a final interview with the applicant, Joanne, for the next day. She has also replied to several emails. Though she still has many emails waiting responses, she plans to get on to these in the last hour of her day. She has blocked in time to write up the plan for the new conference just before lunch, when she will turn off her phone and computer. She plans to work on the marketing idea tomorrow. With her clearer thinking, she can see that one difficult project is plenty for today, and the deadline is not so urgent for this project. It's a great start to the day, the week, and her new position.

Surprises About the Brain

- Conscious thinking involves deeply complex biological interactions in the brain among billions of neurons.
- Every time the brain works on an idea consciously, it uses up a measurable and limited resource.
- Some mental processes take up a lot more energy than others.
- The most important mental processes, such as prioritizing, often take the most effort.

Some Things to Try

- Think of conscious thinking as a precious resource to conserve.
- Prioritize prioritizing, as it's an energy-intensive activity.
- Save mental energy for prioritizing by avoiding other high-energy-consuming conscious activities such as dealing with emails.
- Schedule the most attention-rich tasks when you have a fresh and alert mind.
- Use the brain to interact with information rather than trying to store information, by *creating visuals for* complex ideas and by listing projects.
- Schedule blocks of time for different modes of thinking.

SCENE 2

• • • • • • • •

A Project That Hurts
to Think About

t's 10:30 a.m. Paul picks up a thick pile of papers still warm from the printer. It's a fifty-page brief for a software project that's bigger than anything he's done before. That's the good news. The bad news is that in less than an hour the client is expecting to receive a proposal, in preparation for a lunch meeting today.

Paul had wanted to start working on a proposal when the brief arrived four days ago. At the time he had read over the document lightly, but it had all seemed too complex, and he'd gotten distracted by something else. Because he usually needs only an hour to write up a proposal, he hadn't worried about doing this before today. He hadn't noticed how much bigger the project was than normal.

Paul reads through the document closely. It's now 11:00 a.m. With only thirty minutes to write up the proposal, he finally starts to work on a spreadsheet. He somehow loses ten minutes getting the formulas just right. He senses there are still hours of spreadsheet work ahead of him before he can provide an accurate quote.

The trouble with this proposal is that the project involves too much information for Paul to hold in his mind at once. It hurt his head to think about it last week, which is why he didn't continue working on it then, and it's hurting his head now. It's so complex he doesn't even know where to start. For a few minutes he takes up more space on his already overcrowded stage by adding a new thought: whether he has a problem procrastinating. Then he decides to try to do what he normally does. He starts a spreadsheet and attempts to quickly build a budget for the project, line by line. After a few minutes he can see he has hours to go. He needs a new strategy.

Paul decides to try quickly writing up the general wording for his proposal and to leave the final price to drop in last. He hopes to get some inspiration while he prepares the document. At 11:25 a.m., with five minutes left, he panics and takes a guess at his costs. He makes the guess a little high to be safe, but is worried he might have missed some expenses. Then he puts on a 100 percent markup. Just as he is about to send the proposal, he notices a typo. As he goes to fix this, his computer crashes. Valuable minutes are lost. He emails the proposal five minutes late, hoping the client won't notice. A few minutes later, when he prints a copy, he notices two grammatical errors. Frustrated, he tries to push aside his feelings by preparing to leave for the meeting, but his frustration doesn't dissipate.

As you discovered in scene 1, your capacity to make decisions and solve problems is limited by your energy-hungry prefrontal cortex. Paul has hit a second limit of the prefrontal cortex here: there's a limit to how much information can be held in mind and manipulated at any one time. That's because the stage is small, smaller than generally acknowledged. To make a series of important decisions this morning, Paul has to quickly make sense of a huge amount of information. To do that, he needs to learn to maximize his prefrontal cortex's limited processing space.

THE STAGE IS SMALL

The mental stage is smaller than you might expect. It's more like a stage in a child's bedroom than the one at Carnegie Hall. It can hold only a handful of actors at a time. Put too many on, and others get bumped off. With so little space available, it's easy to get overwhelmed and make mistakes.

So just how much space do you have up there? This question has perplexed scientists for some time. You've probably never heard of George A. Miller, but you may have heard of the outcome of a study he did in 1956. Miller found that the maximum number of items a person can hold in mind at once is seven. The trouble with Miller's research being so well known is that it's wrong, or at least often misinterpreted. This misinterpretation may be the cause of universal angst: many people think they have a problem because they can't hold that much information in mind.

Here's some hope for tortured souls. A wide survey of new research in 2001 by Nelson Cowan, at the University of Missouri–Columbia, found that the number of items you can hold in mind is likely not seven. It's more like four, and even then this depends on the complexity of the four items. Four numbers, no problem. Four long words, and it starts to get harder. Four sentences, unless the sentences are familiar—a memorized prayer or an advertising jingle—are very difficult indeed to keep in mind. And the participants in these studies were all young adults. Think about it. Four sentences. That's not a lot. No wonder meetings often seem so chaotic. No one can make sense of what's going on.

A further clue to this limitation comes from the makeup of any idea you want to hold in mind. It's easy to remember the sequence of words *catch*, *dream*, *ringer*, *Fred*. But try remembering the sequence *thirl*, *frugn*, *sulogz*, *esdo*, four words using the same number of letters from the same alphabet. It's practically impossible to remember four words in a language you don't speak, or in gibberish. The point is, the stage works efficiently when you bring items onto it made up of elements embedded in long-term memory. This also explains why it's hard to think about new ideas unless they connect to existing ideas. Without long-term hardwiring underpinning the meaning of a new concept, you can't bring the concepts onto the stage easily.

It gets worse. A study by Brian McElree at New York University found that the number of chunks of information you can remember accurately with no memory degradation is, remarkably, only one. This study states, "There is clear and compelling evidence of one unit being maintained in focal attention and no direct evidence for more than one item of information extended over time." While you can obviously remember more than one thing at a time, your memory degrades for each item when you hold a lot in mind.

Clearly this is a limitation worthy of respect. Yet for some reason a lot of people want to buck against it. Long-term memory seems so enormous, and isn't the brain the snazziest bit of technology in the known universe? It just doesn't seem right. Consider a scientific anecdote involving a young graduate student who refused to accept the limitation of his working memory. The student locked himself in a soundproof room for days on end to see if he could increase his working memory for audio tones. Unfortunately the only thing that increased was his need for psychotherapy.

There appears to be real limits to the amount of information that can be held in the prefrontal cortex at any one time, though with chunking and other helpful strategies, this can be significantly maximized. But what about when you try to do something with information on the stage, such as make a *decision* between two actors? An entire field of study, called *relational complexity*, explores this question. Relational complexity studies show over and again that the fewer variables you have to hold in mind, the more effective you are at making decisions.

TOO MANY MAPS

To understand why the stage is so small, let's look at Paul's challenge of trying to write his proposal, from his brain's perspective. As Paul reads his client's brief, he tries to hold dozens of variables on his stage at once. The client, a retail chain, is asking for a quote on the design and installation of new software. They want customers to be able to swipe their credit cards upon entering the store, pick up items, and then walk out without stopping to pay. An electronic reader would charge the items to their credit card as they got close to the front door, via a device tagged to each item. (And if there were a problem, a noise would go off.) Paul's project is to design the software for this system and install it in five hundred stores. Paul has done similar work before, which is why the client called him. And the project itself isn't too big; he thinks he can do it. The problem is that the amount of information

Paul needs to hold on his stage to create a quote for this project is too much to hold at once, especially if he has to do this in a short time. He's trying to squeeze thirty actors onto a space designed for four at most. As a result, the play just won't start. This is a challenge a lot of people face at work now—it's not just an avalanche of information, it's that we have to process this information so quickly now, too.

To understand why this is a problem for Paul's stage, let's take one variable: the idea of storing people's credit card details. This concept alone activates a complex map that contains billions of connections across Paul's brain, and not just in his prefrontal cortex. (A map is a similar idea to a network or circuit.) The map for "credit card processing" connects to maps in Paul's language circuits; for example, connecting the word *credit card* with words such as *interest, default,* and *expiration.* The "credit card processing" map connects with long-term memories. It links to the memory of the first credit card Paul had, every credit card he's had since, and the last time his credit card went over the limit. There are also connections to his motor cortex, with a circuit for the movement of getting the card from his wallet, swiping it, and putting it back. (A map rich enough that Paul could literally do this with his eyes closed.) If you could draw the map for "credit card processing" on paper, the map for the brain circuitry involved would be more complex than the street directory for the entire United States.

Once again, what looks simple, upon closer inspection involves massive complexity. Yes, you can hold seven simple numbers in mind, if you're just trying to remember them, as long as you keep repeating the numbers (until the pattern becomes embedded in your longer-term memory), and you are doing this in your native tongue, while doing nothing else at all. What you can't do is bring more than a handful of complex maps onto the stage at one time. It's too much for the brain to manage.

IT IS A COMPETITION

One of the reasons for the prefrontal cortex's space limits comes down to the principle of competition. Holding a complex concept on your stage usually involves activating visual circuitry. When you think, you picture how a concept connects in space with other concepts. (Working memory is always either visuospatial or auditory, and the former is much more efficient.) Visual awareness works in a competitive way. Circuits compete with one another to form the best internal representation of the external object. Robert Desimone, one of the key scientists at MIT's McGovern Institute for Brain Research, discovered that the brain is capable of holding only one representation of a visual object at a time. It's like the well-known optical illusion where you see either a vase or an old woman in the same illustration. The brain must settle on one perception at any moment; you can't see both at once. You can, however, switch between the dominant perceptions at will, which is an intriguing aspect of these illusions.

For Paul, the map for "credit card processing" would activate many of the same submaps needed for other concepts such as "invoicing the client." The brain doesn't like it when circuits are pulled in several directions at once. It doesn't take many items to be activated before you have the same few million circuits trying to be used by different maps. A conflict results.

DOING YOUR BEST WITH A SMALL SPACE

Since there are limitations to the number of concepts that can be held in mind at one time, the fewer you hold in mind at once the better. The ideal number of new ideas to try to comprehend at once seems to be just one. If you have a decision to make, the most efficient number of variables is likely to be two: Should I turn left or right? If you have to hold more information in mind, try to limit ideas to three or four at once.

I like to think of maximizing working memory as like having a tiny studio apartment, but doing creative things to use the space well, such as putting in a bed that folds into the wall, using lots of mirrors, and hanging shelves up high. When you hear about cognitive improvements that come from brain-training games, they are not coming from making the apartment bigger, but from improving the efficiency of subskills—such as getting information on and off the stage more efficiently by *simplifying* and *chunking* more effectively; and getting better at choosing what to put on the stage and what to keep off, which means learning to *choose your actors carefully*. People intuitively use these three techniques all the time. By better understanding these techniques, you may find yourself using them more, because the circuit underpinning the technique is larger and thus easier to find.

SIMPLIFY

Imagine you are working on a computer with limited RAM, (which means it can't hold much short-term information at any one time). You want to create a single-page document with four high-resolution color photographs. Each time you move the photos around, the computer takes several seconds to redraw everything. To get the positioning right, it would be better to create some low-resolution photos and then move those around on the page to see where they sit best. Once the placement is right, you can then insert the high-resolution color pictures. Graphic designers use this technique of "roughing out" concepts all the time. Screenwriters use "storyboards" to describe how a story progresses, with each board a simple cartoon summarizing complex events. The boards can be moved around easier than reorganizing a whole script. Using a less defined representation of an idea frees up resources needed for important functions such as taking different perspectives, adding or taking out elements, or reordering things.

This ability to simplify complicated ideas into their core elements is a habit that most successful business executives have developed. It's often the only way they can make complex decisions. In Hollywood, for example, the ideal pitch for a new movie is supposed to be so short that a studio can "get" it in just one sentence. (There is a story that the movie *Alien* was pitched as "*Jaws* in space"; the pitch uses existing elements that people know well, in high-level summary form, requiring the least possible energy to get the idea onto the stage.) Simple is good; simplest is best. When you reduce complex ideas to just a few concepts, it's far easier to manipulate the concepts in your mind, and in other people's minds. This is simply because the stage is small. If Paul knew just how small, he might have simplified the project as much as possible. He could have reduced the brief to its salient points, perhaps one line for each key issue, to make sense of it. Instead he did the opposite, going into detail with an attempt to build a spreadsheet line by line.

CHUNK

Here's a little experiment. Take ten seconds to memorize this string of ten digits: 3659238362.

What was this like? Could you repeat this string easily? Now memorize a new string of ten digits for ten seconds, 7238115649, but this time do it by chunking the numbers into pairs, for example "seventy-two, thirty-eight," etc.: 72 38 11 56 49.

If you did this experiment with a stopwatch you would notice how much easier it is to memorize the second set. A number of studies, including those by Professor Fernand Gobet at Brunel University in the UK, show that the brain learns complex routines by automatically grouping information into chunks. The size of a chunk roughly relates to the time it takes you to say each item to yourself. For example, it's easier to say "seventy-two, thirty-eight, eleven, fifty-six, forty-nine," than it is to say "seven thousand two hundred and thirty-eight, one thousand one hundred and fifty-six," and so on. The chunks created when you try to memorize four-digit numbers are too big to stay on-stage easily. The key to this is timing: the best chunks take fewer than two seconds to think about or repeat aloud.

A 2005 article called "The Expert Mind," by Philip E. Ross, in *Scientific American Mind* magazine illustrates how chess masters excel at the game. The article argues that chess masters develop names (i.e., chunks), for complete layouts of the board. They might have a chunk for a game where the other player starts and moves the far left pawn one step, and another chunk for when that player moves the same pawn two steps. They have seen how both games evolve so many times that each game has been memorized and can be recalled in a flash. This enables them to compare the two chunks easily. Expert chess players don't think hundreds of moves ahead. They still hold only a few chunks in mind at a time like the rest of us, but these few chunks can each represent a set of dozens of moves. Becoming an expert in any field seems to involve creating large numbers of chunks, which enables you to make faster and better decisions than amateurs. Current thinking is that it takes about ten years of practice to develop sufficient chunks in any new field to achieve "mastery."

When chunking, each of the four items held onstage can represent millions of other bits of information. Imagine you were trying to rethink your life priorities. You could create chunks for "work," "family," "health," and "creativity." It would be far easier to make life changes by reordering these chunks, than by trying to understand and rethink your entire life history and future plans, which is impossible to do with a small stage. Creating chunks allows you to interact with complex patterns not just on a chessboard but in many domains of life, including your internal life.

Chunking would help Paul get his pricing done in time. He could break the project up into fewer than four chunks, and then break these chunks down again until he started to make connections around how specifically to price the project. Three to four chunks seems to be an ideal number to hold in mind at once, with three being the optimum in many situations.

The brain naturally wants to chunk anytime you hit the limits of what can be held on the stage. It's something you do without noticing. As with simplifying, having an explicit understanding of this process rather than just doing it implicitly will help you chunk more often and more efficiently.

CHOOSE YOUR ACTORS CAREFULLY

If Paul's stage can hold roughly only four actors at once, each one of which can be a chunk containing other actors, then the next question becomes, which four actors are going to be the most useful at any moment?

In scene 1, I introduced the idea that it takes more energy to get some actors on the stage than others. Actors often get onto the stage because they are in the front row, not because they are the most *useful* actors for that moment. When Paul first tries to price his project in half an hour, he quickly fills up his stage with details about the project and finds himself frozen, his stage too full to process anything.

Imagine you are running a meeting with six colleagues and have to make a big decision about whether to invest in a new business. The best four items to hold on your stage might be:

1. The organization's overall goals;
2. The desired outcome from the meeting, such as to decide yes or no;
3. The main argument for the investment; and
4. The main argument against the investment.

According to the insights from the first scene, it will be even easier if these four points are not held onstage but are put somewhere for you to see, such as on paper or a whiteboard.

Instead of choosing just the right actors to put onstage, what often occurs in a situation like this is that people's stages become filled with the details of the new business. That's because these details are fresh in mind, easy to get onto the stage. Whereas the issues listed here, while being important, are also a bit intangible; thus they take more effort to consider. We all often think about what's easy to think about, rather than what's right to think about.

How do you choose which are the best actors to have onstage at any moment? From what we have learned about the brain so far, this decision itself takes a lot of energy and a lot of space. So it's best done early, while you have plenty of mental energy, using visuals as well simplifying and chunking. But for now, that's perhaps enough background about the challenges of this terribly limited stage. Let's go back to the story so you can picture what Paul might have done differently if he had understood the space limits of his own prefrontal cortex.

A PROJECT THAT HURTS TO THINK ABOUT, TAKE TWO

It's 10:30 a.m. Paul sits at his desk and stares blankly at the document in his hands. The client is expecting a full quote within the hour. Paul starts to open a spreadsheet to build a budget up from scratch, but a quiet internal voice tells him this will take too long; the process is too detailed. He has learned about simplifying and chunking when dealing with lots of information.

Paul decides to stop and think about a different strategy. To reduce the amount of information being held both in his computer and his prefrontal cortex, he closes down all the computer programs he is working on and opens a new document, a blank sheet. He thinks about what he most needs to keep in mind. He knows he tends to get lost in details easily, which could prevent his getting the quote in on time, so he first writes, "one hour," on the screen, to keep himself focused on getting finished within an hour. He then looks at the project and tries to define what he most needs to achieve, simplifying this goal into one sentence. Initially he gets lost thinking about the coding again, and then tries to focus on his specific objective for this one hour. He comes up with "accurate pricing" as his main objective. Then he tries to define the project itself in one statement. He gets "software for thousands of small transactions." He has simplified the project into its most important points. Now he holds three ideas in mind: "one hour," "accurate pricing," and "software for thousands of small transactions," to see what connections might occur among these ideas.

Holding these ideas in mind, Paul quickly realizes he should break up the task of pricing the project into stages. He then identifies four chunks for the project:

1. Build a detailed project plan.
2. Research existing software versus building from scratch.
3. Write the software.
4. Install.

When he writes up these four chunks, he sees a pattern. He would like to think about all the detail of the software—his brain wants to go there naturally—but he knows he will get lost if he does so. Instead, he inhibits these actors from getting on the stage, and tries to get just one actor up: "Build a detailed project plan." Putting this concept on the stage for a moment is all it takes for him to remember his system for pricing this specific type of work. He remembers that it normally takes a week to get an accurate project plan together with a client, and he knows how to price his time for a week. Next he thinks about the second chunk: "Research existing software versus building from scratch." Holding just this one concept in mind, he remembers how long this kind of work took to do once before.

He sketches out a pricing plan, and then does the same with the next three steps, putting one concept onto his stage at a time. He gets to the third stage, "Write the software," and realizes there is no way to price this stage until he has completed the first two. He decides to illustrate the costs of this stage for two earlier projects that were similar, instead of giving a firm quote. Drawing this schematic saves him hours of calculations based on unknown variables. With the "Installation" chunk, he can calculate from previous installations the time it takes per store, the support time, and so forth. From here he can create a reasonable estimate, which he can put forward with some disclaimers.

Within thirty minutes he has built a simple spreadsheet with a breakdown of the costs. He prints out the document to check for typos, fixes a few, then sends off a final quote fifteen minutes before the deadline. Paul senses the client will be happy to get this material on time, and to see a breakdown rather than just one number. Pleased with the pitch, he has time to catch up on emails before he has to leave.

Consider the two scenarios. In the first scenario, Paul sends his proposal with some typos, after the deadline, and with one figure, which is just a rough guess. This guess could be costly. In Paul's "take two," he sends the proposal early, broken down into logical steps for the client to understand, and with no errors. The financial difference to Paul could be huge. The difference in terms of brain processes is not that big. Paul recognized that the machinery of his brain was not doing the job he wanted, and he changed the functioning of his brain to achieve his goals. This switch, of course, required effort and attention, and asked that Paul understand his own brain patterns and not do what his brain automatically wanted to do. Sometimes seemingly small changes in the brain can have a big impact in the world.

Surprises About the Brain

- The stage is small, much smaller than commonly realized.
- The less you hold in mind at once the better.
- New concepts take up more space on the stage than ideas you know well.
- Memory starts to degrade when you try to hold more than one idea in mind.
- When trying to make a decision between items, the optimal number of items to compare is two.
- The optimal number of different ideas to hold in mind at one time is no more than three or four, ideally three.

Some Things to Try

- Simplify information by approximating and focusing on an idea's salient elements.
- Group information into chunks whenever you have too much information.
- Practice getting your most important actors onstage first, not just the ones that are easiest.

SCENE 3

• • • • • • • •

Juggling Five
Things at Once

t's 11:00 a.m. Emily is walking to a meeting with her senior executives. It's her first meeting with this group, so she gets directions to the meeting room from the assistant at the elevator. On the way down a long hall, her cell phone rings. It's one of the unsuccessful applicants for the job Emily is hiring for today. While trying to let him down nicely, Emily realizes she's lost. She doesn't have a mental map for how the rooms are laid out. She finishes the call, finds her bearings, and gets to the meeting five minutes late, annoyed with herself.

Emily is intelligent, yet she can't follow directions to a room and talk on the phone at the same time. This inability might seem strange based on the ideas about the stage so far, as there are only two items in her attention: "find the room" and "talk on the phone." Why are only two items overwhelming her prefrontal cortex?

As the meeting participants settle in, Emily notices a colleague checking his phone when her own buzzes with an email arriving. She isn't used to being "always on," but she has so much to do she decided she needs to be contactable at all times. The device came with her promotion. She would turn it off, but she's worried about missing something urgent. The email is from Joanne, the woman whom Emily wants to hire as her assistant. Their meeting needs to be rescheduled.

Emily responds straightaway, keeping a partial eye on the meeting. As she types a message, she feels almost a little dizzy, a bit like how she feels when she tries to read as a passenger in a car. Her brain is doing something it doesn't want to do. She finishes and focuses her attention on the meeting. Her phone buzzes again.

It's Joanne, with another question. The same mild nausea arises as Emily types up a quick response.

"Emily?" A voice enters her awareness with a thud. It's the CEO.

"I was just asking if you'd like to introduce yourself to the team."

"Sure." She pauses, feeling disoriented. She stammers a "thank you" for the promotion and says she has big plans for the year. She's worried that people might think she's a flake who can't speak well in public.

Emily is a great presenter, always ready to make a strong impression at a moment's notice. What's thrown her performance here is another limitation of the prefrontal cortex, one that much of humanity wishes didn't exist, especially those with lots to do. Emily has discovered that there is a limit not just to how much information you can hold at once, as Paul found in the last scene, but also to what you can do with that information at any time. Try to push past this limit, and something will give, which tends to be accuracy or quality. With so much to do every day, Emily needs to rewire her brain so she is more efficient at juggling multiple mental tasks without affecting her performance.

THE ACTORS CAN PLAY ONLY ONE ROLE AT A TIME

While you can hold several chunks of information in mind at once, you can't perform more than one conscious process at a time with these chunks without impacting performance. We now have three limitations: the stage takes a lot of energy to run, it can hold only a handful of actors at a time, and these actors can play only one scene at a time.

While it is physically possible sometimes to do several mental tasks at once, accuracy and performance drop off quickly. The consequences can be harsh. An investigation into a fatal train accident showed that the driver sent a text message at the precise moment that the train accidentally sped up while rounding a corner.

Most people have firsthand experience with this limitation. It's easy to drive and chat with a friend on a well-traveled route. Go to a new destination, however, and the conversation slows right down. Drive on the other side of the road in a foreign country and you will need to focus hard to stay on the correct side of the road. Changing the radio station while driving on the other side of the road is almost impossible until the new way of driving is embedded in long-term memory. Likewise, simply changing one letter on your computer keyboard slows down your writing substantially. The brain now has two things to do at once: remember where the keys are and focus on writing.

As I mentioned in scene 1, the main mental processes relevant to getting work done are understanding, deciding, recalling, memorizing, and inhibiting. To understand why the actors can play only one scene at a time, let's explore these processes further.

Understanding a new idea involves creating maps in the prefrontal cortex that represent new, incoming information, and connecting these maps to existing maps in the rest of the brain. It's like holding actors onstage to see if they connect with the audience. Making a *decision* involves activating a series of maps in the prefrontal cortex and making a choice between these maps. It's like holding audience members onstage and deciding between them, as with an audition for a chorus line. *Recalling* involves searching through the billions of maps involved in memory and bringing just the right ones into the prefrontal cortex. *Memorizing* involves holding maps in attention in the prefrontal cortex long enough to embed them in long-term memory. *Inhibiting* involves trying *not* to activate certain maps. It's like keeping some actors off the stage.

Each of these processes involves complex manipulations of billions of neurological circuits. The key here is that you have to finish one operation before the next can begin. The reason is similar to the explanation of why the stage is small: each process uses incredible amounts of energy and many of the same circuits, so it's easy for competition for circuits to occur. It's like using a calculator: you can't multiply and divide two numbers at the same time.

When engaged in conscious activities, your brain works in a serial way: one thing after another. It's a different experience from when you just observe a scene but don't pay much attention, as when Emily was looking for Madelyn for coffee at 9:00 a.m. In that instance her brain was "parallel processing"—taking in multiple streams of data, but not doing much with it.

DUAL-TASK INTERFERENCE

This idea that conscious processes need to be done one at a time has been studied in hundreds of experiments since the 1980s. For example, the scientist Harold Pashler showed that when people do two cognitive tasks at once, their cognitive capacity can drop from that of a Harvard MBA to that of an eight-year-old. It's a phenomenon called *dual-task interference*. In one experiment, Pashler had volunteers press one of two keys on a pad in response to whether a light flashed on the left or right side of a window. One group only did this task over and over. Another group had to define the color of an object at the same time, choosing from among three colors. These are simple variables: left or right, and only three colors. Yet doing two tasks took twice as long, leading to no time saving. This finding held up whether the experiment involved sight or sound, and no matter how much participants practiced. If it didn't matter whether they got the answers right, they could go faster. The lesson is clear: if accuracy is important, don't divide your attention.

Another experiment had volunteers rapidly pressing one of two foot pedals to represent when a high or low tone sounded. This exercise took a lot of attention. When researchers added one more physical task, such as putting a washer on a screw, people could still do it, sort of, with around a 20 percent decrease in performance. Yet when they added a simple mental task to the foot-pedal exercise, such as adding up just two single-digit numbers, (a simple 5 + 3 =), performance fell 50 percent. This experiment revealed that the problem isn't doing two things at once so much as doing two conscious mental tasks at once, unless you are okay with a significant drop in performance. I learned this lesson the hard way recently. I was on a phone call on my airpods and started looking for an item in another room at the same time. The result was jamming my toe under a door, an injury that took weeks to heal.

Despite thirty years of consistent findings about dual-task interference, many people still try to do several things at once. Workers of the world have been told to multitask for years. Linda Stone, a former VP at Microsoft, coined the term *continuous partial attention* in 1998. It's what happens when people's focus is split, continuously. The effect is constant and intense mental exhaustion. As Stone explains it, "To pay continuous partial attention is to keep a top-level item in focus, and constantly scan the periphery in case something more important emerges."

THE IMPACT OF DOING TOO MUCH

A study done at the University of London found that constant emailing and text-messaging reduces mental capability by an average of ten points on an IQ test. It was five points for women, and fifteen points for men. This effect is similar to missing a night's sleep. For men, it's around three times more than the effect of smoking cannabis. While this fact might make an interesting dinner party topic, it's really not that amusing that one of the most common "productivity tools" can make one as dumb as a stoner. (Apologies to technology manufacturers: there are good ways to use this technology, specifically being able to "switch off" for hours at a time.) "Always on" may not be the most productive way to work. One of the reasons for this will become clearer in the chapter on staying cool under pressure; however, in summary, the brain is being forced to be on "alert" far too much. This increases what is known as your *allostatic load*, which is a reading of stress hormones and other factors relating to a sense of threat. The wear and tear from this has an impact. As Stone says, "This always on, anywhere, anytime, anyplace era has created an artificial sense of constant crisis. What happens to mammals in a state of constant crisis is the adrenalized fight-or-flight mechanism kicks in. It's great when tigers are chasing us. How many of those five hundred emails a day is a tiger?"

Despite the depth of scientific research out there about the problems inherent in partial attention, people continue to stretch themselves to do more at the same time even though the benefits they may be receiving are minimal. Being "always on" seems like a logical solution. Ergo, if you have more emails and messages than can be handled in the time at your desk, then do them everywhere. Also, the idea of having access to your work 24/7 is much easier to bring to your stage than an uncertain solution not in your audience at all, such as changing your email habits. The surprise result of being always on is that not only do you get a negative effect on mental performance, but it also tends to increase the total number of emails and messages you get. People notice you respond to issues quickly, so they send you more issues to respond to.

If you push yourself, in the short term being always on can seem as if you're being productive. The cost on the brain, however, can be significant, as Emily discovered when she experienced a dose of dual-task-interference-induced nausea during her meeting. Think about someone asking you a hard question as you try to make a decision about something as simple as what to eat for lunch. You can almost do it, but the effort is exhausting.

What people tend to do is what Emily tried to do in the meeting. They try to hold several focuses at once and switch rapidly between them. You might think this is a fine idea. But consider what happens when you hold tasks in the background. Given your small working memory, you decrease the amount of data that can be held for what you want to focus on at any moment. Instead of four items on the stage at once, you may be down to three, or even just two. Space is being taken up in working memory for the items held just off the stage. Though it's not been studied yet, it's reasonable to assume that the most energy-intensive items are those most likely to be bumped off the stage first. And, worse still, these are most likely to be conceptual items, such as more abstract goals or other more subtle objectives. This tendency could explain why, when your stage is overloaded, it's easy to lose track of your overall intent. The big actors get pushed off the stage first.

Whenever you multitask, and more than one task requires any amount of attention, accuracy goes down. Aside from doing only one thing at a time (which most people who receive two hundred emails and messages a day will simply scoff at), what other options are there?

There are three possible answers to this juggling act dilemma. One solution is to embed or automate more of what you do, which means you *get the audience to do more work*. Another possible solution is to *get information onstage in the best possible order*. The third possible solution is to *mix up your attention*.

GET THE AUDIENCE TO DO MORE WORK

Businesspeople sometimes say they can multitask just fine. It's true that you can be on a conference call and reply to emails at roughly the same time. However, the reality is you are not doing two tasks that use the stage at any one time. You are switching attention between tasks. The result is reduced attention to the conference call, so that key points may be missed and new ideas may not "soak in." Studies by memory researchers also show that for long-term memories to form, close attention has to be paid to information. You might hear the conference call but, afterward, remember little of what was discussed.

There is one possible solution here. You can learn to juggle lots of balls at work the same way a clown learns to juggle lots of balls: practicing specific activities over and over until they become embedded, which means the activities are not being managed by your prefrontal cortex. Once an activity is embedded, you can then add another activity at the same time. You can keep adding more layers as other layers are embedded. An example is learning to drive: You embed holding the wheel. Then you embed using the accelerator and brake, then this, too, becomes automatic, and you can now learn more subtle skills, such as parking.

I have learned the keystrokes on my computer for saving, cutting, pasting, and undoing, all of which I now do with little conscious attention. This allows me to produce better-quality writing in less time, because I don't need conscious resources for regular actions. When you embed a repetitive task, you are pushing routines down into the brain region called the basal ganglia, first mentioned in scene 1.

The basal ganglia (there's several of them) are central to how the brain stores routine functions. These functions are called routines because they are steps that fit together in a certain order, as in a dance. Your basal ganglia recognize, store, and repeat patterns in your environment. The basic operating principle is somewhat like the "if-then" function in software coding. An example might be, "If you pick up a hot drink, don't gulp it straight down; test it with a tiny sip first." This routine is stored in complex maps. Each map contains the instructions for firing millions of nerves to move hundreds of muscles in just the right order, for just the right length of time, and with just the right strength, to pick up a hot drink in a mug with a handle and bring it to your mouth and sip it.

The basal ganglia have a finger in every pie. Data streams in and out from most other parts of the brain via long white-matter connections. White-matter connections are like long-range cabling that joins up different brain regions. The prefrontal cortex is also well connected to other brain regions, while some regions, such as the amygdale, have

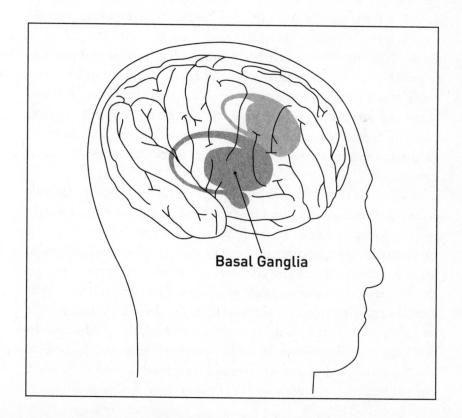

Basal Ganglia

more limited sets of connections to other regions. The well-wired basal ganglia pick up patterns not only in physical movement, but also in light, sound, smell, language, events, ideas, emotions, and in all other perceived stimuli. Next time you unconsciously smell a carton of milk before drinking it, or automatically check to see that you have your credit cards before a lunch meeting, thank your basal ganglia.

The basal ganglia have a big appetite for patterns. One study showed that only three repetitions of a routine is enough to begin the process of what is termed *long-term potentiation*, or what I call here hardwiring. The basal ganglia are also quiet eaters: they pick up patterns without conscious awareness. In a study in Montreal, volunteers in a brain scanner had to press one of four buttons on a keyboard to represent where a light flashed on a screen. The group was divided in two: one group was given a random sequence of patterns; the other, a sequence that repeated. The repeating sequence was complex enough that people couldn't consciously pick it up. However, their basal ganglia did. The group hearing the repeating pattern was 10 percent faster at typing the sequence. After the experiment, both groups were asked to punch in any pattern they had recognized, but neither could do so better than the other. Their basal ganglia had noticed the patterns implicitly, but the volunteers could not explicitly identify them. You might recall similar experiences—for instance, if you drive to a new office one day and somehow seem to "just know" the way the next day. This type of awareness is a subtle knowing. You couldn't describe the route to someone else. A pattern has been formed in your basal ganglia, even though you can't describe it.

The basal ganglia are highly efficient at executing patterns. Use this resource every way you can. Once you repeat a pattern often enough, the basal ganglia can drive the process, freeing up the stage for new functions. Develop routines that can be repeated over and over again: How you call people. How you open up a new document, how you delete emails, how you schedule your time. The more you use a pattern, the less attention you will need to pay to doing this task, and the more you will be able to do at one time. While this process is obviously not possible with higher-order tasks such as writing a complex new proposal, you might be surprised how much can be embedded. For example, I can now, using keystrokes, take less than three seconds (I timed it) and almost no attention to respond to an email with a smiley face, which says essentially, "Got your email, and I am happy."

GET INFORMATION ONSTAGE IN THE BEST POSSIBLE ORDER

Another way to maximize your one-at-a-time attention resources is to get things onto the stage in the best possible order. Imagine trying to decide on a location for a beach holiday in an Airbnb with some friends. Decisions need to be made in a certain order. You can't work out how much food to buy until you know how many people are coming. And you won't know how many people are coming until you choose a date for the vacation. If you went shopping before you had confirmed everyone, you would find yourself thinking in circles, unable to make decisions.

You have probably experienced something similar, perhaps a project at work where you kept going over and over the same decision. This is one of the implications of the serial nature of the prefrontal cortex and conscious mental processing, and it's called a *bottleneck*. A bottleneck is a series of unfinished connections that take up mental energy, forming a *queue*. Other decisions wait in a queue behind the first decision. It's a bit like when a computer printer jams and other documents bank up waiting to print. The printer icon bounces on your screen, sending an "alert" to let you know there's a problem. In a similar way, when a thought keeps recurring, a decision may be holding up other decisions. If you could make a list of thoughts you paid attention to over the course of a week, you would find a range of recurring thoughts. Decisions that get caught in queues, which you try to answer but fail to, are one of the great wasters of a brain's resources.

How do you address issues in a queue? Perhaps a decision higher up needs to be made. If you are decorating a house and can't decide what color to paint the walls, you are probably missing a higher-up decision about the overall color scheme you want. There seems to be a most efficient way, a path of least resistance, to thinking tasks. Taking time to work out the right order to make decisions can save a lot of effort and energy overall, reducing unresolved issues in your *queue*.

Reducing queues stops you from putting the same things on and off the stage over and over again, which gives you more energy, more space for other information, and overall more resources for focusing on other tasks.

MIX UP YOUR ATTENTION

One final technique for dealing with having to juggle is to mix up how you use attention. The idea is similar to what I talked about in scene 1, around scheduling work according to the type of mental task needed. Essentially, if you have to do several things at once, limit the time you spend in the partial attention state. Consciously decide how long you will split your attention, then go back to focusing on one thing. An example of this is leaving your smartphone on for only a limited number of hours a day while you work, perhaps only in the afternoons, when you are not trying to do focused work.

It can be helpful to let others around you know that you are splitting your attention. It's distracting to have to try to establish if someone is listening in a meeting or not. When running a conference call, it might be helpful to be explicit about who is paying 100 percent attention and who is doing other things. When a topic comes up that requires full attention from a specific person, that person can be alerted that their full attention is required.

With all this in mind, let's now take a look at what Emily might have done differently if she had understood this limitation of her brain.

JUGGLING FIVE THINGS AT ONCE, TAKE TWO

It's 11:00 a.m. Emily is due at a meeting with all the senior executives, her first with this group. She gets directions to the room from the assistant and heads down the hall. Her cell phone rings. She knows she can focus on only one thing at a time, and needs to pay attention to where she is going. She switches the call to voice mail and arrives at the meeting on time.

During the meeting, Emily notices someone checking his phone and then hears her own buzzing quietly. She knows if she starts to answer emails, she will lose the thread of the meeting discussion. She asks a question of the group about the agenda for the meeting, so she can make a conscious decision as to whether to split her focus. She learns she will be introducing herself in a few minutes and decides to switch her phone to airplane mode. Emily knows that speaking to the group will need her full attention. For the ten minutes before she has to present herself, she focuses on each person in the room for a moment to get a sense of who he or she is. As she focuses on them, she starts to feel more connected to them and more at ease. She remembers a prior meeting with a couple of them, and the good conversation they had; she is animating a rich network for each person, which will be useful to have in her working memory when she goes to talk. She makes a note to send one of them an invitation to meet for coffee. By the time her introduction comes around, she feels alert but calm.

During her introduction, she comes across as strong and confident. She peppers her talk with insights she remembers from the meeting with the two colleagues in the room, and they are impressed with her memory for detail. After her presentation she lets people know she's going to check messages for three minutes and then switch off again. She starts to read a more detailed email, then gets disoriented and decides to focus on the meeting. She switches back to airplane mode so she's not tempted to respond to any emails. Toward the end of the meeting, there are ten minutes of discussion not directly relevant to her. She uses the time to delete some emails without trying to do two things at once.

Surprises About the Brain

- You can focus on only one conscious task at a time.
- Switching between tasks uses energy; if you do this a lot you can make more mistakes.
- If you do multiple conscious tasks at once you will experience a big drop-off in accuracy or performance.
- The only way to do two mental tasks quickly, if accuracy is important, is doing one of them at a time.

- Multitasking can be done easily if you are executing embedded routines.

Some Things to Try

- Catch yourself trying to do two things at once and slow down instead.
- Embed repetitive tasks where you can.
- Get decisions and thinking processes into the right order to reduce "queues" of decisions.
- If you have to multitask, combine active thinking tasks only with automatic, embedded routines.

SCENE 4

• • • • • • • •

Saying No
to Distractions

It's 11:30 a.m. Paul is meeting with his potential client for lunch in an hour's time. Before then he wants to work out what resources he will need if he wins the credit card project. He has sent off his proposal, but hasn't yet worked out some of the details: who should be on the team, how he would structure the team, and the time line for delivering the project. While he is confident he can do the job, his basal ganglia detect a pattern. Though he can't put words to it, there is something bugging him, a subtle, weak connection from deep in his brain. While he can't pinpoint it right now, it's a memory of needing to be better prepared. It's probably an experience long forgotten, of meeting with a client without being fully prepared, and experiencing strong emotions as a result. The brain remembers a feeling connected with a situation long after the details can't be easily recalled.

Paul gets a blank piece of paper and tries to sketch out which of his suppliers might be best involved in the project. A vague image of an old supplier starts to emerge. Just then, a telemarketer calls. It takes time to find out what she is selling and to get her off the line, as Paul isn't comfortable being rude. Unfortunately, interacting with the telemarketer takes energy, too, something he doesn't have enough of right now. Five minutes later he is still staring at the blank sheet of paper when a gentle ping announces the arrival of new emails. For a moment he thinks he should ignore them, but ignoring them takes effort, too. The first email is from Eric, one of his suppliers, with a question about their school project. Paul and Eric are upgrading the computers for the school their children both attend. It takes Paul ten minutes to respond. He gets tense at the distraction, which he takes out on Eric, with terse answers to his questions.

Paul finishes his email to Eric and tries to start thinking about the project again. Each time he restarts, it takes more effort to focus, and he has a smaller pool of energy reserves to draw on. With each change of focus, Paul needs to get the current actors off the stage and new ones on. The old actors may keep jumping back up because they are just in the front row, which requires *inhibition*. All this takes a lot of energy, something Paul is low on by this time of the morning.

Paul goes to the refrigerator for a snack. Staring at last night's leftovers, he remembers what he was thinking about before the email assault and goes back to his computer. He tries to find the supplier that popped into his consciousness earlier. A moment later he thinks about tonight's neighborhood game of poker and drifts into thinking about last week's game, too; he wished he'd not taken so much money to the game; he knows that if he's not winning he will spend everything he brings. His attention comes back to the present. He notices his computer desktop is messy and starts to put documents into folders. Along the way he notices a project he had forgotten about and opens this file. The phone rings. It's Emily. She has a few minutes downtime and wants to talk about a project she is working on. Paul is torn between talking to her and getting ready for his meeting. Emily misreads Paul's response as disinterest. She tells him she needs his support at the start of her new job, and he responds that he is really busy, too. He suddenly looks at his watch. It's already time to leave.

Despite the importance of the thinking Paul intended to do, he couldn't get started amid all the distractions. His mind wanders everywhere except where he wants it to go. To be more effective at work, he needs to learn to manage distractions better, both external and internal ones. He needs to change his brain so that he can focus more effectively when it's important to him.

EXTERNAL DISTRACTIONS

Distractions are everywhere. And with the always-on technologies of today, they take a heavy toll on productivity. One study found that office distractions eat up an average 2.1 hours a day. Another study found that employees spend an average of 11 minutes on a project before being distracted. After an interruption, it takes them 25 minutes to return to the original task, if they do at all. People switch activities every 3 minutes, either making a call, speaking with someone in their cubicle, or working on a document.

Microsoft has a division that studies the way people work, to develop efficiency-improving software. (According to Microsoft, if you're looking for a technological solution to being more efficient, getting a bigger computer screen is one of the few clear winners.) To reduce the impact of interruptions, they are testing out different techniques, such as more subtle "alerts" (for example, changing the color of a screen object). The challenge is that any distraction, however small, diverts your attention. It then takes effort to shift your attention back to where it was before the distraction, especially when a circuit is new or weak. Each time Paul tries to start planning for this project, he has to reactivate billions of brand new, fragile circuits, circuits that can disappear in a moment like a piece of lint in the air.

Distractions are not just frustrating; they can be exhausting. By the time you get back to where you were, your ability to stay focused goes down even further, as you have even less glucose available now. Change focus ten times an hour (one study showed people in offices did so as often as twenty times an hour), and your productive thinking time is only a fraction of what's possible. Less energy equals less capacity to *understand, decide, recall, memorize,* and *inhibit*. The result could be mistakes on important tasks. Or distractions can cause you to forget good ideas and lose valuable insights. Having a great idea and not being able to remember it can be frustrating, like an itch you can't scratch, yet another distraction to manage.

Part of the solution involves managing *external* distractions: instant message alerts, emails beeping, phones ringing, people walking into your office. Once you understand how much energy is involved in high-level thinking such as planning and creating, you might be more vigilant about allowing distractions to steal your attention. One of the most effective distraction-management techniques is simple: switch off all communication devices during any deeper thinking work. Your brain prefers to focus on things right in front of you. It takes less effort. If you are trying to focus on a subtle mental thread, allowing yourself to be distracted is like stopping pain to enjoy a mild pleasure: it's too hard to resist! Blocking out external distractions altogether, especially if you get a lot of them, seems to be one of the best strategies for improving mental performance.

One particularly important issue is how you use your smartphone. Recent studies show that your smartphone can impact your ability to think well and affect your IQ, even if it is off but still in the room. For your phone to have no measurable effect on you, it has to be off and out of sight in another room.

INTERNAL DISTRACTIONS

A lot of the distractions we all deal with, however, are not external; they are *internal*. As adolescence hits and people become more conscious of an inner life, many notice that their minds are hard to control. Strange thoughts pop into awareness at odd moments. The mind likes to wander, like a young puppy sniffing around here and there. As frustrating as this tendency can be, it's normal. One reason for your wandering attention is that the nervous system is constantly processing, reconfiguring, and reconnecting the trillions of connections in your brain each moment. The term for this is *ambient neural activity*. If you were to look at the electrical activity even in a resting brain, it would look like planet Earth from space with electrical storms lighting up different regions several times a second. The result is a stream of thoughts and images emerging into conscious awareness. A similar process occurs as you dream, when neural connections form behind the curtain of awareness and emerge into the mind. This constant connecting happens when you are awake, too, but most of your hundreds of thoughts each minute never get much attention and disappear into the background. It's like random audience members jumping onto the stage, getting their two seconds of fame, and then leaving. It's easy to be distracted by these unwanted actors if you are not alert. There's some evidence that schizophrenia involves this kind of interruption— an inability to inhibit these task-irrelevant signals that most of us are able to dampen down and effectively ignore.

It's a good thing that random thoughts quickly disappear, as it's hard enough to stay focused even without intrusions. One study showed that people on average hold a thought for only ten seconds before flitting off to something else. The actors are easily distracted, like a theater troupe that leaves the stage every few minutes just because it's a nice day outside or someone sneezed, or for no reason whatsoever. Unless you put in some effort to keep them onstage, it's hard to get a scene completed.

Trey Hedden and John Gabrieli, two neuroscientists from MIT, studied what happens in the brain when people are distracted by internal thoughts when doing difficult tasks. They found that lapses in attention impair performance, independent of what the task is, and that these lapses in attention involve activating the medial prefrontal cortex. The medial prefrontal cortex is located within the prefrontal cortex itself, around the middle of your forehead. It activates when you think about yourself and other people. This region of the brain is also part of what is called the *default network*. This network becomes active when you are not doing much at all, such as in between any focused mental activities. Hedden and Gabrieli found that when you lose external focus, this default brain network activates and your attention goes to more internal signals, such as being more aware of something that may be bothering you. When Paul gets distracted by the thought of last week's poker game, he loses the thread of finding the supplier, and he doesn't get back to this thought until it's too late.

For centuries, philosophers have written about the difficulties of controlling the mind. One famous metaphor from Eastern philosophy involves the "Elephant and the rider," where the conscious will, the rider, tries to control the larger and uncontrollable unconscious mind, the elephant. With the prefrontal cortex taking up just 4 percent of total brain volume, modern brain science seems to affirm the truth of this metaphor. The prefrontal cortex, central to conscious decision-making, has a degree of influence, but the rest of the brain is bigger and stronger.

This points to the importance of increasing the strength of the networks linking the prefrontal cortex with the rest of the brain.

DRIVEN TO DISTRACTION

The big problem with distractions, whether internal or external, is that they are, well, distracting. That's not just because focusing takes effort, as I mention earlier. Being distracted by new information around you is also a "knee-jerk" reflex action, as much as, say, a knee jerk. One theory about why this is the case is that over millions of years, your brain learned to orient attention to anything unusual. Or, as scientist and philosopher Jonathan Haidt at New York University says, we are the descendants of people who paid a lot of attention when there was a rustle in the bushes. A new shape of car, a flash of

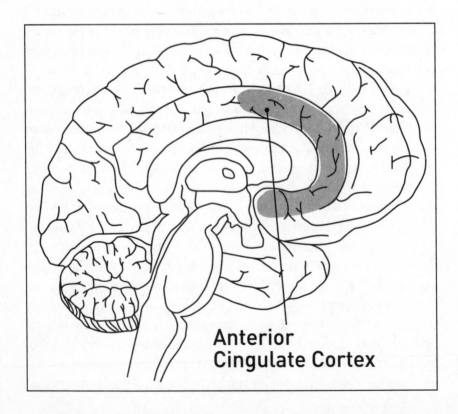

**Anterior
Cingulate Cortex**

light, an odd sound underfoot, or a strange smell—all get our attention because they stand out, because they are *novel*.

The brain region important for detecting novelty is called the anterior cingulated cortex (see diagram, page 50). It's thought of as your error-detection circuit, because it lights up when you notice something contrary to what is expected, such as when you make a mistake or feel pain. This quirk of nature is harnessed by all forms of marketing and advertising, as well as by people seeking to meet someone of the opposite sex. Novelty gets attention. In small doses, novelty is positive, but if the error-detection circuitry fires too often, it brings on a state of anxiety or fear. This partly explains humanity's universal resistance to wide-scale change: big changes have too much novelty.

There are many distractions at work, as Paul found during his morning. There are the external distractions, the emails and phone calls, the folders that need filing. Then there are the internal distractions, such as the recollection of the poker game. Some internal distractions may be generated by the limitations of the stage itself. There simply might not be enough glucose available for intensive thinking, so you keep losing your train of thought. You might be trying to hold too much information in mind, more than four concepts at once, and so you keep losing items. Or there might be other decisions in your "queue," earlier decisions that need to be made that keep jumping into view. Or there could be things in your short-term memory that are taking up space, that are just not helpful and need to be pushed aside. Perhaps now you can begin to see why Arnsten calls the prefrontal cortex the Goldilocks of the brain. Everything has to be just right for it to work well.

DRIVING AWAY FROM DISTRACTIONS

With all this possibility for chaos on the stage, you might wonder how you ever stay focused. Human beings have developed specific neural circuitry for this process, though it doesn't work the way you might expect. Maintaining good focus on a thought occurs through not so much how you focus, but rather how you inhibit the wrong things from coming into focus.

A common test that neuroscientists use to study the act of focusing is the "stroop" test. Volunteers are given words printed in different colors, and are asked to read out the color of the text, not the word itself. In the example here, the brain has a strong desire to answer, "Gray," for option C, as it's easier for the brain to read a word than to identify a color.

a. **Black**

b. **Gray**

c. **Gray**

d. **Black**

To not read the word *gray* requires *inhibition* of an automatic response. Using scanning technologies such as functional magnetic imagery, which records changes in blood flow in the brain, neuroscientists have observed people inhibiting their natural responses, and discovered the brain networks that are activated when this happens. One specific region within the prefrontal cortex keeps showing up as being central for all types of inhibition. It's called the ventrolateral prefrontal cortex (VLPFC), and it sits just behind the right and left temples. The VLPFC inhibits many types of responses. When you inhibit a motor response, a cognitive response, or an emotional response, this region becomes active. It appears that the brain has many different "accelerators," with different parts of the brain involved in language, emotions, movement, and memories. Yet there is a system used for all types of braking, the VLPFC (while other brain regions are involved in braking as well, the VLPFC appears to be central). Your ability to use this braking system well seems to correlate closely to how well you can focus.

PUTTING ON THE BRAKES

The fact that the VLPFC sits within the prefrontal cortex has some big implications. If you were a car company and were building a new type of on-road vehicle, you would make sure the braking system was made out of the most robust materials possible, because brake failure is not a happy thing. Well, in the case of human brains, the opposite has happened. Our braking system is part of the most fragile, temperamental,

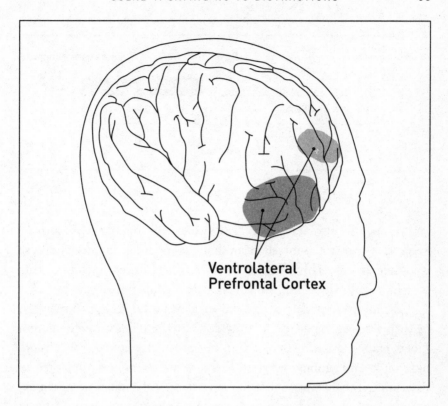

**Ventrolateral
Prefrontal Cortex**

and energy-hungry region of the brain. Because of this, your braking system works at its best only every now and then. If cars were built like this, you'd never survive your first drive to the store. All this makes sense when you consider it: stopping yourself from acting on an urge is something you can do sometimes, but is often not that easy. Not thinking about an annoying, intrusive idea at times can be very difficult. And staying focused—well, sometimes that appears downright impossible.

One poignant implication of your braking system being located in the prefrontal cortex is that your capacity to put on the brakes decreases each time you do so. It's like having a car whose brakes pads nearly disappear each time you apply them, unless there is a long rest between uses. Roy Baumeister, from Florida University, a scientist introduced in scene 1, set up a situation where people had to resist eating chocolate while alone in a room. He found that those who resisted the chocolate gave up more quickly on a difficult task afterward. "Self-control is a limited resource," says Baumeister. "After exhibiting self-

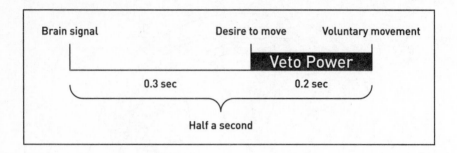

control, people have a reduced ability to exhibit further self-control." Each time you stop yourself from doing something, the next impulse is harder to stop. This tendency explains a lot, including why dieting is so hard and why I eat so much chocolate while writing.

Let's look a little deeper into the science of inhibition, as it seems to be such a central capacity. A study done in 1983 by the late Benjamin Libet, from the University of California–San Francisco, sheds more light on what is going on here. Libet and his colleagues attempted to determine if there was such a thing as "free will." They set up an experiment that enabled them to understand the timing involved when people decided to do a voluntary activity, in this case, to lift up a finger. What they found is that half a second before a "voluntary" movement, the brain sends a signal called an *action potential*, which relates to a movement about to occur. This action potential occurs a long time, in neuroscience terms, before any conscious awareness of the desire to move the finger. The brain decides "I will move my finger now" about 0.3 seconds before you are aware of it. When you get the nerve to talk to the attractive person across the room, your brain was being bold three-tenths of a second before you.

Once you become aware of a desire to move something—be it your finger in an experiment or yourself across the room to take your chances—your brain made this decision millions of connections ago. After this point, there are 0.2 seconds during which you are aware of being about to move, but haven't yet taken the action. This 0.2-second window is a decent amount of time, long enough for the mind, with some practice, to notice an urge and perhaps intervene.

This is a big point. You don't have much ability to intervene in the signals sent out by your brain. With all that ambient neural activity, the brain sends out all sorts of crazy ideas into the mind. But you do have "veto power," the ability to choose whether to act on an impulse. However, without an awareness of the separation of these processes—"brain signal, desire, movement"—it's likely you will go directly from brain signal to movement, the way most other animals do. You need to be able to discern these small time scales. The way to do so is by paying attention to your mental experience and noticing urges for action as they unfold.

It seems that you may not have much free will, but you do have "free won't" (a term coined by Dr. Jeffrey M. Schwartz), which is the ability to avoid urges. However, you have only a small window in which to inhibit a response. And, of course, if your stage is too full, you may not have the space to hold the concept of inhibition there. It's starting to become clear why, when you're tired, hungry, or anxious, it's easier to make mistakes and harder to inhibit the wrong impulses.

TIMING IS OF THE ESSENCE

Inhibiting distractions is a core skill for staying focused. To inhibit distractions, you need to be aware of your internal mental process and catch the wrong impulses before they take hold. It turns out that, like the old saying goes, timing is everything. Once you take an action, an energetic loop commences that makes it harder to stop that action. Many activities have built-in rewards, in the form of increased arousal that holds your attention. Once you download or just look at your emails and notice messages from people you know, it's so much harder to stop yourself from reading them. Most motor or mental acts also generate their own momentum. Decide to get out of your chair, and the relevant brain regions, as well as dozens of muscles, are all activated. Blood starts pumping and energy moves around. To stop getting out of your chair once you start will take more veto power and more effort, than to decide not to get up when you first have the urge. To avoid distractions, it's helpful to get into the habit of vetoing behaviors early, quickly, and often, well before they take over.

There's something interesting in the timing of all this, and to help this make sense I want to revisit the experiment of the 1980s mentioned in the last scene. Two groups of people copied a complex pattern of lights that shone in front of them, typing the pattern into a matching keyboard. One group had a random pattern. The other group had a complex but repeating pattern that couldn't easily be discerned consciously. The people given the repeating pattern somehow typed 10 percent faster. Their unconscious mind, most likely the basal ganglia, had picked up the pattern and was anticipating the next flash of light, even though they couldn't consciously identify the pattern in tests afterward.

Here's where that experiment gets more interesting. Some of the time the participants could identify the pattern. They could explain it in words or type it out. These people could type the sequence 30 to 50 percent faster than if there was no pattern. The people who knew the pattern consciously were also able to execute this pattern within 0.3-second intervals. Three tenths of a second is very close to the gap between noticing you want to take an action and taking an action, as we learned from the Libet experiment.

When you develop language that describes an activity, at least in this experiment, it's more likely that you can catch yourself about to do something before taking the action. Having explicit language gives you more veto power. When you have words for a pattern, which means the prefrontal cortex is involved, a lot more is possible in relation to that pattern.

This finding about language is relevant to managing distractions, but it's also relevant to everything we have talked about so far. If you have language for the way your mental stage gets tired, you will catch this exhaustion as it happens. If you have language to describe the feeling of having too much on your stage at once, you will be more likely to notice it. In some ways this whole book is about helping you develop explicit language maps within the prefrontal cortex for experiences that until now have occurred only *implicitly*. This book can help make your brain's processes more explicit and, as a result, give you more veto power over dealing with too much information, too many demands on your attention, too many distractions, and other challenges to be explored in coming scenes.

The brain is easily distracted, and distractions have a big energy cost. Staying focused requires learning not just to switch off your smartphone and put it in the other room when you need to get big things done. The harder part is learning to inhibit impulses as they arise. To inhibit impulses, you must veto them before they turn from impulse into action. And you are more likely to be able to veto an action if you have explicit language for the mental processes involved. It pays to know a lot about how your brain works, so you can catch your brain while you try to work.

Before we get too abstract about all this, let's bring this to life in a more tangible way by going back to Paul to see what he might have done differently if he had been better at managing distractions in his own brain.

SAYING NO TO DISTRACTIONS, TAKE TWO

It's 11:30 a.m. Paul is meeting his potential client in an hour at a restaurant across town. Between now and then he wants to think through what resources he will need if he wins the credit-card project. He senses he needs to consider the details beyond just the pricing before he meets the client.

Paul gets out a blank piece of paper and tries to sketch out which of his suppliers might be best for the project. A vague image of a supplier he worked with some time ago starts to emerge into his awareness. Just then a telemarketer calls, and Paul accidentally answers his cell phone without thinking, as his braking system is not well resourced while he's focused on the project. This distraction reminds him that he won't complete this delicate but energy-hungry job of planning the project if he has to deal with distractions. While he tries to get off the phone, he uses embedded motor routines to switch off his computer and all other phones in the room.

Once he finishes the call and switches off the phone, Paul starts to think about the project again. He feels clearer knowing there won't be other distractions, that a part of his stage has been freed up that would otherwise be paying subtle attention to whether the phone might ring. With the stage cleared, Paul remembers where his thoughts were before the call. He reactivates a complex but fragile network of billions of neurons. The supplier he was trying to remember comes to mind. He emails the supplier, who's available for a quick conversation and is keen to work on the project. He gets back on the phone just for this call and together they map out a plan for how the project might unfold. Speaking about ideas activates more circuits than merely thinking about those same ideas, which makes it easier to stay focused: the network is more robust.

Paul is relieved to be prepared before the meeting. He turns on his computer and creates and prints out a basic plan, which will make him look more organized. Looking at his watch, he sees he has a few minutes to spare. The phone rings. It's Emily. She has a few minutes downtime after her meeting and wants to talk about her first day in her new position. He tells her she will do just great, and she thanks him for the support. They have been talking about the kids for a while when Paul looks at his watch. It's time to leave for his meeting.

Surprises About the Brain

- Attention is easily distracted.
- When we get distracted it's often a result of thinking about ourselves, which activates the default network in the brain.
- A constant storm of electrical activity takes place in the brain.
- Distractions exhaust the prefrontal cortex's limited resources.
- Being "always on" (connected to others via technology) can drop your IQ significantly, as much as losing a night's sleep.
- Focus occurs partly through the inhibition of distractions.
- The brain has a common braking system for all types of braking.
- Inhibition uses a lot of energy because the braking system is part of the prefrontal cortex.
- Each time you inhibit something, your ability to inhibit again is reduced.
- Inhibition requires catching an impulse when it first emerges, before the momentum of an action takes over.
- Having explicit language for mental patterns gives you a greater ability to stop patterns emerging early on, before they take over.

Some Things to Try

- When you need to focus, remove all external distractions completely.
- Reduce the likelihood of internal distractions by clearing your mind before embarking on difficult tasks.
- Improve your mental braking system by practicing any type of braking, including physical acts.
- Inhibit distractions early before they take on momentum.

SCENE 5

• • • • • • • •

Searching for the Zone of Peak Performance

Paul gets into his car to go meet with the potential client. The meeting is being held over lunch, at a restaurant a thirty-minute drive away, in a part of town Paul doesn't visit often. As he pulls onto the road, he thinks about not having to deal with emails or calls for thirty minutes, and lets out an audible sigh of relief. Ten minutes down the road, entering the freeway, he realizes he is driving in the wrong direction. He has taken the route he takes each day to drop his daughter off at school.

He senses he is going to be late, and this anxiety heightens his alertness. He starts thinking hard about the journey. Realizing he is about to hit midday traffic, he works out a route through the back streets to save time. He gets off the freeway and starts zigzagging through smaller streets, a little extra acceleration underfoot. Driving this way takes a lot of focus. Five minutes before the meeting time, he starts to get tense and remembers a time he missed a meeting. This internal distraction makes him miss a turn and lose more time. Finally he turns another corner and sees the restaurant just ahead. He walks into the restaurant one minute past the hour. As the host takes him to his table, Paul notices his colleagues are already halfway through a cup of coffee, and appear far more relaxed than he feels.

* * *

During his journey to the lunch meeting, Paul experienced the full range of performance of the prefrontal cortex, from *under-arousal*, where he made a mistake, to the *right level of arousal*, where he performed well, to *over-arousal*, when he fell apart again. What Paul experienced illustrates the last significant limitation of the prefrontal cortex: that it's fussy. The prefrontal cortex needs just the right level of arousal to make decisions and solve problems well. For Paul to focus, he needs to learn not just to reduce distractions as we saw in the last scene, but also how to get his brain into the right level of arousal.

THE ACTORS ARE HIGH-MAINTENANCE

Arousal in any region of the brain means its level of activity. Neuroscientists can measure the levels of arousal in any brain region in several ways. One way is via an electroencephalogram (EEG), which measures types and levels of electrical activity in the brain with the placement of sensor pads on the skull. Another way to measure arousal is through increased blood flow, commonly measured by an fMRI, functional magnetic resonance imagery, which uses powerful magnets to read changes in blood levels due to iron in the blood.

In the brain there's a constant shifting of arousal. As some regions become busier, others go quiet. It's like watching a city from high above and seeing millions of people pour from the suburbs and into the center in the morning, then head back out at the end of the day. This isn't a bad metaphor for a brain at work, as for much of the workday, blood, oxygen, nutrients, and electrical activity pour into the prefrontal cortex to support the intense activities it is being called on to undertake.

A certain level of arousal is required for the prefrontal cortex to work at its best. That level is quite high, but not *too* high. Not only are the actors on your mental stage easily distracted, they are also high-maintenance. They need the perfect amount of pressure to perform at their best. Too little pressure—no audience, for example—and they don't focus. Too much, and they forget their lines.

YOU AND INVERTED *U*

Researchers have known for one hundred years that there is a "sweet spot" for peak performance. In 1908, scientists Robert Yerkes and John Dodson discovered a fact about human performance that they called the inverted *U*. They found that performance was poor at low levels of stress, hit a sweet spot at reasonable levels of stress, and tapered off under high stress. The verb *stress* means "to emphasize," and it's not necessarily a negative thing. It's wrong to think your performance would improve if stress disappeared from your life. It takes a certain amount of stress just to get out of bed in the morning. This type of stress is known as eustress, or positive stress. Positive stress helps focus your attention.

When Paul first drove off, he suffered from the rare phenomenon of being too happy on the job. He was feeling so good he forgot to create and hold an image in mind of where he was going. When you don't activate the prefrontal cortex, you tend to operate on habit, and your basal ganglia take over. At that moment, Paul was at the bottom left hand of the inverted *U*: without enough stress to perform well. It's how you might forget a conference call scheduled while on a summer holiday: your capacity to remember important tasks melts away with the warm sun and piña coladas. You're too relaxed.

When Paul started to focus on navigating the back streets, he was in the "sweet spot" of the *U*, with everything working at its best. More stress resulted in better performance. The fear of being late increased his attention to the task at hand. Many people feel they can't focus unless there is a tight deadline. A small amount of fear or a sense of urgency can definitely generate a helpful level of focus at times. (Though, as you will see in the next scene, this is more useful in situations where physical or routine tasks are needed.)

Toward the journey's end, when Paul thought he might be late, he started to panic and missed a turn. He wasn't able to focus on the map, the one in his mind or in his hand. His arousal levels impacted his performance negatively. Unfortunately this is the mental state that millions of workers find themselves in every day. Too much arousal reduces performance above a certain level.

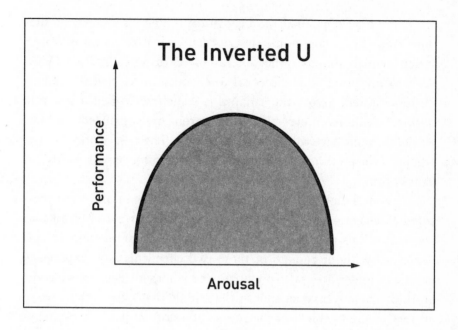

IF THE CHEMISTRY IS RIGHT

Some exciting new research illustrates the underlying physiology occurring at various levels of arousal. Many of these discoveries come
from Amy Arnsten, the neurobiologist from Yale. Arnsten has spent
twenty years studying the prefrontal cortex, down to the level of neurons, synapses, neurotransmitters, and even genes. Her findings help
explain why the prefrontal cortex is so fussy, and point the way to
techniques for managing states of arousal.

First some background. Neurons, the nerve cells of the brain, are
not directly connected to other neurons. Instead, there is a small gap
between them called a synapse. An electrical signal travels down a
neuron cell body and is converted into a chemical signal at the synapse. There are receptors on both sides of the synapse that receive
messages from these chemical signals. Synapses send and receive one
of two signals: either what is called an *excitatory* signal, which tells
the neuron to do more of something, or an *inhibitory* signal, which
tells it to do less of something. Excitatory signals are part of what
is called the behavioral activation system, or BAS. Inhibitory signals

are part of what's called the behavioral inhibition system, or BIS. (Fun fact: this is commonly described by scientists as the BIS/BAS system, which sounds rather like a cute cartoon character.) Okay, back to my point: this electrical-to-chemical-to-electrical communication system across the synapse is sometimes called a synaptic "firing." Trillions of ever-changing neurons are organized into networks through patterns of neuronal firing. These networks are the "maps" I keep talking about, such as the map for "credit cards" in Paul's brain.

Arnsten discovered that whether a synapse in the prefrontal cortex fires correctly depends on having just the right levels of two neurochemicals present. These chemicals are dopamine and norepinephrine. Without enough of these two chemicals, you experience boredom, under-arousal. Too much, and you experience stress, over-arousal. There is a sweet spot in the middle that's just right. "We're all very aware of this over the course of a normal day," Arnsten explains. "For example, when we've yet to wake, or are tired at the end of the day, it's very hard to get organized, or do any complex prefrontal cortex activity. Then when you're too stressed, you get massive levels of norepinephrine and dopamine, and this causes all networks to disconnect, and leads to shutting off of nerve firing altogether. We end up with nerve cells saying very little to each other." For the prefrontal cortex to function well, the brain must deliver just the right levels of these two neurochemicals to incredibly large numbers of constantly changing connections. No wonder focusing can seem so difficult at times.

Your brain chemistry changes across the course of a day as a result of natural environmental stimulation. If you almost walk in front of a bus, you will probably find yourself more awake for the rest of the afternoon. If you go to the forest after a stressful day, you will probably feel calmer. However, you can also shift your own chemical states through various mental techniques, without risking your life or needing to find some trees. These techniques can help you either decrease or increase levels of alertness or interest, or both.

GOING ON ALERT

If you have ever had to run a conference call soon after waking up in the morning, you will know that "arousal" is important for making sense of the world. Putting aside stimulants such as caffeine (which is, like a bigger computer screen, a proven technique for increasing mental performance), there are two main strategies for increasing arousal.

One strategy, perhaps the easiest and quickest, is to increase adrenaline levels by bringing "urgency" to a task. Norepinephrine, also known as noradrenaline, is the brain equivalent of the adrenaline most people feel before public speaking. It's the chemistry of fear. When you are scared, you pay intense attention; you are highly alert. Fear brings a deep and immediate alertness. Norepinephrine also turns out to be important for binding circuits together in the prefrontal cortex.

You can play various "tricks" on yourself to generate the release of this chemical. Visualizing an activity generates a similar metabolic response to actually doing it. One study found that picturing yourself doing a finger exercise increased muscle mass by 22 percent, which was close to the 30 percent achieved by doing the exercise for real. (For those thinking this sounds too good to be true, remember that you still have to put in the effort, a lot of effort, to keep mentally focused on doing the exercise.)

If your alertness is too low, you can generate adrenaline by imagining something going wrong in the future, literally visualizing a scary event. In the previous scene, Paul was under-aroused because it was Monday morning, before lunch. It was hard for him to focus. Even small distractions took over. In this situation he could have used his brain to imagine himself standing, unprepared, in front of the client. The resulting fear would have increased his levels of norepinephrine, which would have helped him focus. A professional boxer once explained to me the secret of his success. He used to imagine that going into the ring could kill him, which would make him train like a maniac. I use a similar trick when writing. If I can't focus, I imagine posting a blog online and many people finding mistakes in it. That wakes me right up.

The key to this technique is not to let the imagery take on a life of its own. You want to arouse the brain just enough to get motivated, but not so much that you end up obsessing about your fear and increasing your allostatic load.

GETTING INTERESTED

Another way Paul could have put his brain into the right neurochemical gear is through the dopamine pathway. Where norepinephrine is the chemistry of *alertness*, dopamine is the chemistry of *interest*. Good levels of both chemicals are required to generate the right level of arousal, but each chemical has a different impact on its own.

Dopamine is released in a number of situations. First, the dopamine level rises when the orbital frontal cortex detects novelty, something unexpected or new. Children love anything new. The chemical rush from novelty goes from interest to an intense desire in a flash. Humor is all about creating unexpected connections. Watching funny film clips or telling jokes increases dopamine levels. If you have ever noticed that saying something the first time is easier than repeating it, you are noticing the pleasant buzz of new circuits being activated for the first time. Each time you say the same thing afterward takes more effort, as you no longer have the dopamine buzz of novelty.

Paul could improve his focus by changing simple aspects of how he works. Just changing the height of his chair could be enough of a fresh perspective to release more dopamine. Or he could speak out loud about his project to someone, allowing him to get a novel perspective again. Or he could listen to some jokes, call a friend he likes to have a laugh with, or just read something interesting and entertaining.

Scientists have also found that expecting a positive event, anything the brain perceives as a reward, generates dopamine. Rewards to the brain include food, sex, money, and positive social interactions. So Paul could have put his prefrontal cortex into the right neurochemical sweet spot by focusing on the possible rewards of his doing a great job on this proposal, the money he could win, and the future rewards that would come his way.

Looking across all the research, one finds that there may be advantages to using positive expectations or humor to generate arousal, rather than fear. Humor and positive expectations activate both dopamine and adrenaline. Fear yields adrenaline, but the expectation of negative events reduces dopamine. Fear also activates other chemicals that can have a negative impact on your body over time.

TOO MUCH AROUSAL IS NOT A GOOD THING

Over-arousal may be more of a serious problem than under-arousal. In a study of 2,600 British workers, one half had seen a colleague reduced to tears because of pressure, and more than 80 percent had been bullied during their careers. People everywhere are experiencing information overload, which involves too much stimulation from too many ideas at once. Paul experienced the dark side of over-arousal when he missed a turn on the way to the meeting, which resulted in panic.

Over-arousal means there is too much electrical activity in the prefrontal cortex. To reduce this arousal, you might need to reduce the volume and speed of information flowing through your mind. When you can't seem to think, writing ideas down to get them "out of your head" can help. If your stage doesn't have to hold this information, there is less activity overall.

Another strategy involves activating other large regions of the brain, which tends to deactivate the prefrontal cortex. One example is to focus your attention on the sounds around you, which activates brain regions involved in perceiving information coming into the senses. You could also activate the motor cortex, by doing anything physical, such as taking a walk, which makes oxygen and glucose flow to more activated areas of the brain such as the motor cortex. If one brain region is overly activated, you can sometimes solve this problem by activating another. This is a long way of saying, "take a walk when you're stressed," but it's also helpful to understand why this works.

Too much arousal involves not only experiences such as fear or anxiety. It can also refer to more positive arousal, such as excitement or lust. New lovers tend to "lose their minds" and do all sorts of crazy things in the heat of the moment. One study showed that new lovers' brains have a lot in common with people on cocaine. Dopamine is sometimes called the "drug of desire." Too much dopamine, from being "high with excitement," can also be exhausting. Any activity that triggers the thought of possible rewards can captivate our attention through increased dopamine levels. This is part of how gambling works. It's also how we get addicted to certain apps on our smartphones. Any bit of novel information—an unexpected news story, a video of a cat doing something you thought was physically impossible for cats to do—and you get a lovely rush of dopamine into the brain. This can feel great, but over time we can get addicted in ways that can literally reduce our IQ.

AROUSAL IS INDIVIDUAL

The point at which something is either stressful or engaging varies significantly between people. For one person, riding a bicycle on a bike path in the suburbs might hardly raise his arousal. He needs to put on roller blades and dodge Manhattan traffic to feel focused. For another, the thought of riding a bicycle on a deserted wide pathway might be totally overwhelming. These differences are partly based on previous experiences and other factors we will discuss in the next act. There is a genetic component to this that, while interesting, isn't that helpful for our understanding here. However, there is also a gender component to this inverted *U* that explains many everyday phenomena.

One reason Paul is in trouble this morning is that he left writing up his proposal for the last minute. The client had sent him the brief four days ago, but at the time Paul felt he couldn't focus on it, that things weren't "urgent" enough. Arnsten explains this generally masculine phenomenon: "Estrogen promotes the stress response. This is now me describing the story of my lab—the women get everything done a week ahead of time, as they don't want the pressure, the increased arousal, of the deadline. The men wait till the last minute so they have enough dopamine and norepinephrine to actually push them to finish."

GETTING THE AROUSAL JUST RIGHT

We have explored the experience of over- and under-arousal, but what about the experience at the top of the U, the sweet spot? Hungarian scientist Dr. Mihaly Csikszentmihalyi—for the non-Hungarians, his last name is pronounced "Cheeks-sent-me-high"—has been studying this state for decades. In his 1990 book, *Flow: The Psychology of Optimal Experience*, Csikszentmihalyi describes the experience at the top of the U as being the optimal state between too much stress (over-arousal) and boredom (under-arousal). It's when you are immersed in an experience and time seems to stand still. Paul experienced "flow" when he decided to focus and take the back streets, before he had the fear that he would still be late.

Everyone craves this experience of being in "flow," as it is so energizing. Dr. Martin Seligman, the founder of the positive psychology field, thinks the flow state is one of the three main drivers of human happiness, more important than hedonic happiness, the joy we get from a good meal or fine wine. According to Seligman, flow also seems to involve using your "strengths," a set of behaviors you have become so good at they have become embedded.

I have a theory for why the flow state is so engaging and energizing. Imagine doing something that uses deeply embedded routines that requires minimal effort or attention, such as driving your car. Now think about using these routines to do something slightly different, but something a little harder than normal—something you could do well only if you focused. For example, instead of driving your normal car, drive a rally car, on a race track. Some of the basic skills are there, such as steering and changing gears, but you would need to pay a lot of attention because a few variables are new. What happens is a large number of new connections are made, but from a base of safety, because you already have many connections to build on. The result is a strong flow of dopamine and norepinephrine—without a lot of effort. This flow of neurochemicals occurs because many new connections are forming. The chemistry helps you focus, and this focus then helps creates more new connections. A positive spiral is created where you feel focused and energized.

In summary, the prefrontal cortex is fussy. To function at its peak, it needs just the right levels of two neurochemicals, at just the right point within billions of circuits. These chemicals relate to being alert and interested. Fortunately, as you've seen, there are ways you can intervene in this process to make yourself more or less alert or interested. To make this all a bit clearer, let's explore what Paul could have done differently if he had understood the discoveries about the brain in this scene.

SEARCHING FOR THE ZONE
OF PEAK PERFORMANCE, TAKE TWO

Paul gets into his car to go to the client meeting, a thirty-minute drive away, in a part of town he doesn't visit often. He likes the idea of taking a drive, and relaxes as the car warms up, breathing a sigh of relief at not having to deal with emails for thirty minutes. He knows he needs to focus to get to this part of town, and lifts his alertness to a higher level by picturing himself arriving at the meeting. His adrenaline levels go up. He is about to start to drive when he hears a voice telling him to check the map first. His basal ganglia have seen this pattern before, and in his alert but not overwhelmed state, Paul notices quiet internal signals such as this. He scans the map and works out the best route. He asks the car to turn on one of his favorite songs. Every ten minutes he lowers the volume on the music and checks the map to be sure he's on track. He is focused yet relaxed. In this optimal state, without consciously choosing to, he finds himself using the time to mentally rehearse how to present himself to the client. He remembers to start by asking lots of questions, and to present his other big projects first. He runs through the presentation in his mind, imagining how he will introduce each section of the proposal and what the client might say. All this activity makes him feel alert, focused, and prepared. He arrives a few minutes before the meeting is to begin, with enough time to settle in with a coffee and prepare his documents.

Surprises About the Brain

- Peak mental performance requires just the right level of stress, not minimal stress.
- Peak mental performance occurs when you have intermediate levels of two important neurotransmitters, norepinephrine and dopamine, which relate to alertness and interest.
- You can consciously manipulate your levels of norepinephrine and dopamine in many ways, to improve your alertness or interest.

Some Things to Try

- Practice being aware of your levels of alertness and interest throughout the day.
- Bring your adrenaline level up when needed with a small dose of visualizing a mild fear.
- Bring your dopamine level up when needed, using novelty in any form, including changing perspective, humor, or expecting something positive.
- Bring your dopamine or adrenaline level down by activating other regions of the brain other than the prefrontal cortex.

SCENE 6

• • • • • • • •

Getting Past
a Roadblock

t's noon. Emily has given herself only thirty minutes to write a simple proposal for the new conference she plans to pitch at lunch. She has discovered two things about her brain over the years: that she can bring ideas to her stage with less effort if she writes when a deadline is close, and that writing seems to expand to fill the time available.

After a few minutes, as Emily gets close to finishing her proposal, she has a small insight: she should go to her lunch meeting with a suggested brand name for the conference. The novelty of this insight sparks her interest, pushing up her dopamine levels. But she quickly gets annoyed with herself for not having thought of this before, remembering it can take days to work out the branding for a conference, not a few minutes. Her anxiety level goes up, a little too much for allowing her to think clearly. She stops to reflect for a moment. She decides, despite a strong interest in the branding exercise, to finish the general plan first so she can have a clearer mind to work on the brand. Emily knows her own mind enough to see that a few minutes of an empty stage can generate more ideas than a longer time with distractions.

Emily finishes the general plan and now has ten minutes to come up with a name for the conference. She still doesn't feel she is in the

right state of mind for such creative work: it's pre-lunch and her glucose levels are low. So she switches off her phone and hangs a Do Not
Disturb sign on her door. She knows she can't afford even one distraction with her brain in this fragile state. She puts aside other papers on
her desk, a physical act that also helps mentally clear her stage. Then
she opens a new document on her computer and starts to brainstorm.

She immediately connects to the obvious words about the event—
sustainable business—and starts to think of how to use these words
to make up a name. These words have front-row seats in her audience,
because they have been mentioned a lot lately. It's well established that
you remember, without knowing why, words or concepts you have
seen recently and that these automatically influence your actions, subconsciously. It's a quirk of the brain called *priming*.

Her list begins: "Sustenance," "Sustain," "Sustaining Business,"
"Sustain All," "Sustaining Profits," "Sustain for Gain." She doesn't
like any of these and tries to think of another path to take, but her
mind is locked into this way of thinking. She starts to get distracted.
Not making the connections she wants, her dopamine level goes down,
which makes distractions harder to keep out. She chooses to veto her
attention from taking the path of being annoyed with herself. Instead,
she concentrates her attention on picturing herself presenting the ideas
at lunch, to increase her focus. A few moments later she finds another
run of words on the "sustain" theme. As she heads off to her meeting,
she is pleased that she had the foresight to write the general proposal
first. At least she has a complete proposal, and some possible brand
names to present, though she knows she hasn't got the right name yet.

Emily has followed most of the principles in this book so far. She
schedules work when it is easier to get the actors onstage, she clears
her mind to reduce the amount of information she has to hold, and
she does one thing at a time. She reduces external distractions, and
she vetoes internal distractions. Yet she has still hit an impasse. She
isn't able to find a good conference brand name she wants using just
the conscious mental processes of her prefrontal cortex. She needs to
bring other brain resources to bear. Emily is discovering another surprising finding about the prefrontal cortex. Sometimes the prefrontal

cortex itself is the problem. This is especially the case in creative situations. Emily needs to understand her brain better, to know when and how to switch off its conscious, linear processes, so she can be more creative on demand.

INSIGHTS ARE THE ENGINE OF THE ECONOMY

Emily has hit what is known in neuroscience circles as an *impasse*. An impasse is a roadblock to a desired mental path. It's a connection you want to make but can't. An impasse can be anything from trying to remember an old friend's name, to working out what you will name your child, to suffering full-blown writer's block. While impasses are something we all experience regularly, they are especially relevant when you need to be creative. Being creative involves getting around impasses.

According to Professor Richard Florida, author of *The Rise of the Creative Class*, more than 50 percent of workers today do creative work. They write, invent, design, draw, color, frame, or tinker with the world in some way. Creative people are all about putting together information in a novel way. Novelty gets attention. And in the business world, attention tends to generate revenue. In this way, the creative process is a big engine of wealth creation.

While a little novelty on the one hand can generate a positive dopamine response, too much novelty can be frightening. Put this concept together with the fact that people's inverted Us are so different and you begin to see why new products get such widely varying reactions from the public. (Walt Disney is reported to have said that if he tested a new idea out and people were unanimously against it, he knew he might be on to something.) Most creativity isn't of the *Fantasia* type, but rather slight changes on existing themes. Fifty percent of workers are tinkering at the edges, trying to make existing products or services more interesting. These people hit a lot of impasses.

Consider the other 50 percent of workers who are not "creative." Whether you work at a bank, make sandwiches, manage a currency exchange, or captain a tourist yacht in the Bahamas, you probably spend most of your day executing codified routines stored in your basal ganglia. Then, suddenly, you hit a new problem that makes you think: you're out of mayonnaise, the currencies are whacko (the U.S. dollar is worth *what*?), or the boat's low on fuel. Some problems are solved easily: the sandwich-making manual tells you where to buy more mayonnaise in an emergency. With other problems, you use a mental search function, comparing the problem to previous problems to find a possible solution. On the boat in the Bahamas, you remember what you did when you ran out of fuel once before. You rationed supplies, made the alcohol free, and sailed to any port downwind.

However, with the amount of change today in how business is done, "noncreative" people increasingly run into brand-new problems, problems with no procedures to follow, no obvious answers, and where solutions from similar situations don't work. For example, what's the rule for reducing the production cost of a product you don't understand, one that is manufactured in China, serviced from India, delivered into Europe, and managed by people who have never met? What's needed here is not a logical solution, but one that recombines knowledge (the maps in your brain) in a whole new way. And that's called an *insight*.

Whether you're a creative person tinkering with the shape of a product, or a captain of a ship, knowing how to get from an impasse to an insight can make a big difference to your success. One of the fascinating aspects of the insight experience is how much you need to switch off your stage to have one. In many cases, an overactive prefrontal cortex can be the cause of the roadblock itself.

GOING UNCONSCIOUS

For a long time, insights were thought of as mysterious events that seemed to happen all by themselves. No one knew much about how they worked biologically, so it was hard to develop theories for how to increase them. Today that's less the case, thanks to scientists such as Dr. Mark Beeman.

Beeman is an associate professor at Northwestern University, in Evanston, Illinois. He won't tell you this himself, but Beeman is one of the world's experts on the neuroscience of insights. Beeman is also the kind of person whose high energy requires that you have a strong cup of Java before meeting him just so you can keep up with the conversation.

Beeman's original interest was in how the brain understands language. He was interested in the way we fill in the gaps in language, which opened up another interest: how we solve cognitive problems more generally. This intellectual quest led to a fascination with the insight experience. In 2004, Beeman, along with colleagues John Kounios and others, undertook some groundbreaking neuroscience studies that explored what happens in the brain before, during, and after an insight experience.

"There is a famous old quote from William James on attention: 'Everyone knows what attention is until you try to define it,' " Beeman explains in an interview at his lab. "I think a similar thing could be said about insight. Everybody has insights. It's usually not some great scientific theory; it could just be about how to rearrange the garage so the car actually fits."

In the lab, Beeman studies people having insights as they solve word problems. He believes these simple puzzles have much in common with real-world challenges, which can't be studied easily. A puzzle might involve three words: *tennis*, *strike*, and *same*. The goal is to generate a solution word that has something to do with each of the three words. The solution word for this puzzle is *match*, because you can have a "tennis match," you can "strike a match," and *match* and *same* have similar meanings.

Beeman finds that about 40 percent of the time people solve his problems logically, trying one idea after another until something clicks. The other 60 percent of the time an insight experience occurs. The insight experience is characterized by a lack of logical progression to the solution, but instead a sudden "knowing" regarding the answer. "In insight," Beeman explains, "the solution comes to you suddenly and is surprising, and yet when it comes, you have a great deal of confidence in it. The answer seems obvious once you see it."

See for yourself. Take the words *pine*, *crab*, and *sauce*, and see if you can work out the connecting word that makes sense with each of them. Try to make a mental note of the process that you use to solve the problem. Do you logically work it out? Does it come to you in a flash? When you get the answer, do you "know" it's correct straight-away?

The fact that an insight feels obvious and certain when you see it is a clue to what might be going on in the brain when you have one. Beeman and his team tried to work out if the brain was processing the problem below the level of conscious awareness. According to re-search on *priming*, when someone is told an answer to a problem that their subconscious has already solved, they read that answer faster. Beeman found this was the case. (This is the "a-duh" experience, a term coined by Jonathan Schooler, from the University of California, Santa Barbara, for when someone else tells you the answer to a prob-lem you were working on. The "a-duh" experience is different from the more positive "aha!" experience, when you solve a problem with insight yourself.)

When insights occur, they seem to involve unconscious processing. That makes sense from experience—insights often come from nowhere and at the most unusual times, when you are putting in no conscious effort to solve a problem—such as in the shower, at the gym, or driv-ing on the freeway. This knowledge about insights provides a possible strategy for increasing creativity: let your subconscious brain solve the problem. And when you take a walk in the middle of the workday with your phone switched to airplane mode, you now have hard science to use to explain this when your boss looks at you funny.

Fortunately, there are some more sophisticated strategies emerging for increasing insights, other than just taking a stroll. To understand these, let's dive deeper into the discoveries about the "aha!" moment. (In case you didn't solve it, the answer to the *pine-crab-sauce* puzzle is *apple*. You can have a pineapple, a crab apple, and applesauce.)

STUCK AT AN IMPASSE

It's rather counterintuitive, but scientists have found that one of the best ways to understand insight is to understand what happens just before an insight occurs: the impasse experience. One of the scien-tists leading this research is Dr. Stellan Ohlsson, at the University of

Illinois at Chicago. Ohlsson explains how when facing a new problem, people apply strategies that worked in prior experiences. This works well if a new problem is similar to an old problem. However, in many situations this is not the case, and the solution from the past gets in the way, stopping better solutions from arising. The incorrect strategy becomes the source of the impasse.

Emily is at an impasse when she gets stuck in the loop of words relating to *sustainability*. She has been caught in one way of thinking. Ohlsson's research shows that people have to stop themselves from thinking along one path before they can find a new idea. "The projection of prior experience has to be actively suppressed and inhibited," Ohlsson explains. "This is surprising, as we tend to think that inhibition is a bad thing, that it will lower your creativity. But as long as your prior approach is most dominant, has the highest level of activation, you will get more refined variations of the same approach, but nothing genuinely new comes to the fore." Here is this concept of *inhibition* from scene 4 arising again. The ability to stop oneself from thinking something is central to creativity.

You now have an additional excuse for taking that stroll in the park when you're stuck on a problem. I can just picture someone's last words to their boss before being fired: "I am going for a walk to forget about work and get totally unconscious." As funny as this sounds, it's what the research shows is needed when you get stuck at an impasse. The wrong answers are stopping the right ones from emerging.

Here's an opportunity for a personal experience of the impasse phenomenon. It's a word puzzle that is exceptionally obvious when you see it, yet nearly everyone hits an impasse when they try to solve it. The puzzle goes like this: What does the array of letters H, I, J, K, L, M, N, O stand for? Take a moment to try to answer this, but also make a note of what strategies you try and where you get stuck. How did you do?

A common impasse is to try to solve this puzzle as an acronym ("He Is Just Kindly Laughing" or some such). But the real answer is less contrived and obvious once you see it. What do these letters stand for? Well, they are the letters in the alphabet from H to O. Get it yet? It stands for something you drink every day: H_2O.

This exercise illustrates how challenging it can be to break out of fixed ways of thinking. When you assume that an acronym is the answer, this assumption pushes out other possible solutions. The map for "acronyms" is active in your brain, and the electrical activity holding this in place inhibits other circuits from easily forming. Getting around an impasse is like trying to change the direction of traffic on a bridge: you have to stop the traffic from going one way before it can go the other.

Ohlsson's principle of inhibition explains why insights come in the shower or the swimming pool. It's nothing to do with the water. When you take a break from a problem, your active ways of thinking diminish. This seems to work even at the level of a few moments. Try an experiment: next time you're working on a crossword or other word game on your smartphone, when you get stuck, do something totally different for a few seconds (anything as simple as tying your shoes or stretching; the main thing is not to think about the problem). Then come back to the problem and see what happens. I predict you may notice how sometimes the prefrontal cortex, your conscious processing capacity, is itself the problem. Get it out of the way, and the solution appears.

This quirk of the brain also explains why other people can often see answers to your problem that you can't. Others are not locked into your way of thinking. (I talk about this idea further in a framework called the "clarity of distance" in my book *Quiet Leadership*.) Knowing a problem too well can be the reason you can't find a solution. Sometimes we need a fresh perspective. This is an unusual concept, as normally we think the best person to solve a problem is the one who knows everything about it. With so many impasses each day at work, perhaps what's needed are more thinking partnerships, where one person has a lot of detail and the other very little. Together they can come up with solutions faster than either can on his own.

Let's go back to Emily. She needed to be creative on demand, but got stuck at an impasse, despite doing all the right things to clear her mind first. What should she have done differently? She shouldn't have focused harder on the problem in her last few minutes. She should have done something counterintuitive: taken one of her precious minutes to do something else entirely, something interesting and even fun, to see if an insight popped up. As bizarre as this strategy might seem, Beeman has shown that focusing more intensely, as Emily tried to do by increasing her anxiety through picturing herself at the meeting, doesn't increase insights. It decreases them.

DISTANT CONNECTIONS

Other than taking that career-risking stroll, what else can you do to have more insights? Beeman's research provides clues. He found that people who solved a problem with insight had more activation of a brain region called the right anterior temporal lobe, a region underneath the right ear. This area allows you to pull together distantly related information. It is part of the right hemisphere, the hemisphere more related to holistic connections. Jonathan Schooler showed that when people focus on the details of a scene instead of the big picture, they disrupt the insight process, by shifting their brain into left-hemisphere mode.

Beeman has found that people having insights experience an intriguing brain signal just before the insight occurs. The brain in some regions goes quiet, like a car going into idle. According to Beeman, "About a second and a half before people solved the problem with insight, they had this sudden and prolonged increase in alpha band activity over the right occipital lobe, the region that processes visual information coming into the brain." The alpha activity disappeared exactly at the moment of insight. Beeman continues: "We think the alpha activity signifies people sort of had an inkling that they were getting close to solving the problem, that they had some fragile, weak activation that was hinting at the solution somewhere in the brain. They wanted to shut down or attenuate the visual input so they could decrease the noise in their brain, in order to allow them to see the solution better. Kind of like saying, 'Shut up. I am thinking about something.'" You do this all the time, probably without noticing. You are talking to someone. Then, just for a moment, you avert your eyes, perhaps by looking up, to be less distracted. It's the brain's way of shutting down inputs to focus on subtle internal signals. If you don't do this, insight is unlikely to occur.

Beeman has also found a strong correlation between emotional states and insight. Increasing happiness increases the likelihood of insight, while increasing anxiety decreases the likelihood of insight. This relates to your ability to perceive subtle signals. When you are anxious, there is greater baseline activation and more overall electrical activity, which makes it harder for you to perceive subtle signals. There's too much noise for you to hear well. This is why companies such as Google create work environments that allow for fun and play. They have seen that this increases the quality of ideas.

Other experiments have shown that brain regions involved in cognitive control, which switch your train of thought, are activated prior to insight. You have been thinking about a problem one way, but you now need to switch and think about it in another way, to increase your chances of solving the problem. Just prior to insight, the medial prefrontal cortex tends to become active. This is part of your default network, and it relates to being aware of your own experience. When trying to solve problems while in a brain scanner in the lab, people who had less medial prefrontal activation, but showed greater activation of visual areas in the brain, tended not to have insights. They were looking closely at the problem, but they were not aware of how they were looking. Beeman reached a point in his experiments where he could pick who was most likely to have an insight and who not, before the experiment even started, based solely on their brain-activation patterns.

Here's what Beeman found. People who have more insights don't have better vision, they are not more determined to find a solution, they don't focus harder on the problem, and they are not necessarily geniuses. The "insight machines," those whom Beeman can pick based on brain scans before an experiment, are those who have more awareness of their internal experience. They can observe their own thinking, and thus can change how they think. These people have better cognitive control and thus can access a quieter mind on demand.

These intriguing findings have big implications for training and education across the board. So much of the emphasis at school, university, and in the workplace is on cognition and general intelligence. There is little focus on knowing yourself or on cognitive control. If getting around impasses in the future is going to be important—I can certainly think of a few possible impasses that need solving—then we might need to rethink how we teach problem-solving.

PAY ATTENTION TO YOUR INNER ARIA

With all this research put together, in theory it should be possible to develop techniques and practices that increase insight. I spent more than ten years working on this challenge, which resulted in the development of the ARIA model. ARIA stands for Awareness, Reflection, Insight, and Action. The model both describes the stages of an insight, so you can recognize the process in real time, and provides practical techniques for increasing the likelihood of insight.

Awareness is a state in which the brain focuses lightly on an impasse. In the awareness state, you want to put the problem on the stage, but ensure it takes up as little space as possible so that other actors can get on. To minimize activation of the prefrontal cortex, don't focus too hard, quiet the mind of other thoughts, and simplify the problem as much as possible. A good way to simplify a problem is to describe it in as few words as possible. Saying to yourself, "I want more energy," creates a fair bit less activation in the brain than saying, "I want more energy to focus more on my work and family and make time for exercise and fun."

In the *reflection* phase, you hold the impasse in mind, but reflect on your thinking processes, rather than on the content of thoughts. In the H_2O example, an insight is more likely to emerge if you notice that none of your strategies is working, and then allow totally new strategies to emerge into awareness. The objective is to see your impasse from a high level, not to get detailed. This activates right hemisphere regions that are important for insight, and allows loose connections to form. You also want to activate the easy, unfocused mental state that occurs as you drift awake in the morning, when ideas dreamily flow into mind.

The *insight* stage is fascinating. At the moment of insight, there is a burst of gamma band brain waves. These are the fastest brain waves, representing a group of neurons firing in unison, forty times a second. The gamma frequency signifies brain regions communicating with one another. People in deep meditation have a lot of gamma waves. People with learning difficulties have fewer gamma waves, and someone who is unconscious has almost none.

Here is the gamma band burst in a graph from Beeman. The first

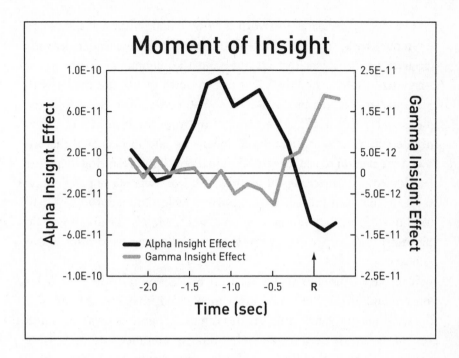

spike in the darker line is the alpha wave, the brain going quiet. The second spike is the gamma band wave that occurs right at the moment of insight.

Insights also come with an energetic punch. You can see it on people's faces, hear it in their voices, and see it in their body language. You can even sense it during a telephone call. It's obvious when you know what to listen for. An insight is a moment when things change. Insights also bring a rush of adrenaline and dopamine. Insights are exciting; they hold your attention and make you feel great.

The *action* phase is your opportunity to harness the energy released upon the formation of an insight. This energy is powerful but short-lived. Think of the high you feel when at the end of a good book the plot comes together and all these strange puzzles suddenly make sense. For a couple of minutes you feel this wonderful sensation, but ten minutes later this has significantly decreased in intensity. While this "high" is present, people will be more courageous and motivated to commit to certain actions, but once the neurochemical cocktail wears off, their motivation will decrease fast.

The ARIA model points to just how valuable brain insights can be. In one workshop I ran, more than seventy business leaders learned about the neuroscience of insights and the techniques for bringing others to insight. They then had five minutes to use the model with one another, on real business challenges. Seventy-five percent of the impasses the business leaders worked on were resolved in just a five-minute conversation. ("Resolved" means an insight occurred that allowed the person to see the situation in a new light, resulting in a clear decision to do something differently.) All I did was show people how to create the right brain state in another person to increase the likelihood of insight. Our brains love an insight. Mostly it's about getting the prefrontal cortex out of the way, and allowing deeper signals to be heard.

The ARIA model can be used on oneself or on others. It helps you remember the brain processes involved in insights: activating a quieter stage and generating greater internal cognitive awareness and control. It can be used to remember an acquaintance's name, solve a crossword clue, or find the next idea for your screenplay. Let's see how Emily might have been more creative on demand with these findings in mind.

GETTING PAST A ROADBLOCK, TAKE TWO

It's noon. Emily has thirty minutes to write a proposal for the new conference. A few minutes into writing, she has an insight: she should go to her lunch meeting with a possible brand name for the conference. She feels her dopamine level go up. She knows an insight heightens the chemistry needed for more insights, so she quickly tries to harness this energy. She turns off all her phones and pagers and hangs a Do Not Disturb sign on her door. She opens a new document on her computer and starts to brainstorm.

Emily connects to the key words in the brief—*sustainable business*—and starts to think of how to use these words to come up with a name. After finding ten words on this theme, she stops and notices the thinking path she has taken. She sees that she has become stuck in the "sustain" theme. She quiets her mind, trying to listen for other threads to follow. She hears a subtle thought, something about the "future," and follows this path. Another ten words. She listens for more clues and soon connects to the idea of insurance, of reducing risk. Another dozen words. As she finishes these, she notices no new themes coming to mind. She knows she has to focus elsewhere to let subtle connections form. She recognizes she is at an impasse, and will probably just keep coming up with more solutions based on these three themes.

She switches her brain out of gear to inhibit the current solutions. She calls Paul to ask about his day, and they chat for a few minutes. In the midst of hearing about Paul's upcoming pitch and how tense he is, a new theme jumps into her mind: "relax." She gets off the phone and comes up with "Relax into the future," and "A relaxed future," then feels she's at a dead end. She switches her attention to looking at pictures of her kids to reduce anxiety. Suddenly she feels a sense of excitement beneath her awareness, and in pops a strong idea: "Future-proof your business." She does a quick search and, finding that the phrase has not been used, rewrites her proposal based on this theme. The additional dopamine pushes her into a state of flow, where she produces some of her best work. She still has time for a rough stab at the other proposals. In her energized state, she crafts more good ideas than she had expected, and heads off to the meeting in a positive state of mind.

Right now you might be encountering an impasse yourself. So far the theme of this book has been about using your prefrontal cortex more efficiently. To be effective at work I have proposed that you need to get a minimum of actors on the stage, in the right order, a few at a time, with the right level of arousal. Yet now I am suggesting that sometimes you have to get everyone off the stage, so that unconscious processes can solve the problem. But *when* and *how* do you decide that it's time to shut down your stage? And the big question, of course, is, *who* exactly is doing all the deciding here? To address these questions, let's take a break from the main story and explore some deeper findings about the brain.

Surprises About the Brain

- It's astonishingly easy to get stuck on the same small set of solutions to a problem, called the impasse phenomenon.
- Resolving an impasse requires letting the brain idle, reducing activation of the wrong answers.
- Having insights involves hearing subtle signals and allowing loose connections to be made. This requires a quiet mind, with minimal electrical activity.
- Insights occur more frequently the more relaxed and happy you are.
- The right hemisphere, which involves the connections between information more than specific data, contributes strongly to insight.

Some Things to Try When You Hit a Mental Wall

- Take the pressure off yourself, get an extension on your deadline, do something fun, reduce your anxiety any way you can.
- Take a break and do something light and interesting, to see if an answer emerges.
- Try quieting your mind and see what is there in the more subtle connections.
- Focus on the connections between information rather than drilling down into a problem; look at patterns and links from a high level rather than getting detailed.
- Simplify problems to their salient features; allow yourself to reflect from a high level, watch for the tickle of subtle connections preceding insight, and stop and focus on insights when they occur.

Intermission:
Meet the Director

It's time for an intermission. Let's take a break from Paul and Emily's story and consider some of the deeper insights emerging about the brain. My proposition so far has been that understanding your brain increases your effectiveness at work. This happens because with knowledge of your brain, you make different decisions moment to moment.

However, just having more knowledge of you brain alone may not be enough. Note the phrases in italics from the last scene with Emily: "She *sees* that she has become stuck in the "sustain" theme. She *quiets* her mind, trying to *listen* for other threads to follow. She *hears* a subtle thought, something about the "future," and *follows* this path." Emily is paying attention to her mental processes as they occur. She is an observer of her own brain at work. Without this act of observation, knowledge of her brain might not change much. Peak mental performance requires a combination of the two—knowing your brain, and being able to observe your brain processes occurring in real time.

In the stage metaphor, the actors represent conscious information. The audience members represent information in your brain below conscious awareness, such as memories and habits. Then there is a character I am calling your *director*. The director is a metaphor for the part of your awareness that can stand outside of experience. This director can watch the show that is your mental life, and therefore your life, make decisions about how your brain will respond, and even sometimes alter the script.

The Director Through History

This idea of a director goes by many names and has been of great interest to scientists, philosophers, artists, and mystics for centuries. At the dawn of Western philosophy, Socrates said, "The unexamined life is not worth living." Today, some people refer to the experience of observing yourself as self-awareness or mindfulness. Sometimes it is called metacognition, which means "thinking about your thinking." Or meta-awareness, which means "awareness of your awareness." Whatever it's called, this phenomenon is a central thread in much of the world's literature, appearing as a core idea in philosophy, psychology, ethics, leadership, management, education, learning, training, parenting, dieting, sports, and self-improvement. It's hard to read anything about human experience without someone saying that "knowing yourself" is the first step toward any kind of change.

With the prevalence of this idea, one of two things is happening here. Perhaps authors are all terrible plagiarists. Or maybe there is something important, universal, and therefore biological about being able to step outside and observe your moment-to-moment experience. Research is pointing to the latter.

Cognitive scientists first recognized in the 1970s that working memory, the stage, had an aspect that they called the executive function. This executive function, in a sense, sits "above" your other working-memory functions, monitoring your thinking and choosing how best to allocate resources. Research into this phenomenon deepened with the development of new technologies in the 1990s, and specifically around 2007 with the emergence of a field called social cognitive and affective neuroscience, sometimes called social cognitive neuroscience.

Social cognitive neuroscience is a hybrid of cognitive neuroscience, the study of brain functioning, and social psychology, the study of how people get along. Before social cognitive neuroscience, neuroscientists tended to focus on how a single brain functioned. Social cognitive neuroscientists study the way brains interact with other brains, exploring issues such as competition and cooperation, empathy, fairness, social pain, and self-knowledge. This last area is of special interest here. Many of the brain regions your brain uses to understand other people are the same as those used for understanding yourself. Social cognitive neuroscientists, excited to explore some philosophically challenging topics, want to get to know this elusive director.

Kevin Ochsner is the head of the Social Cognitive Neuroscience Laboratory at Columbia University in New York City, and one of the two founding fathers of social cognitive neuroscience. As he sees it, "Self-awareness is the capacity to step outside your own skin and look at yourself with as close to an objective eye as you possibly can. In many cases it means having a third-person perspective on yourself: imagine seeing yourself through the eyes of another individual. In this interaction it would be me becoming the camera, looking at myself, observing what my answer was. Becoming self-aware, having a meta-perspective on ourselves, is really like interacting with another person. This is a fundamental thing that social neuroscience is trying to understand."

Without this ability to stand outside your experience, without self-awareness, you would have little ability to moderate and direct your behavior moment to moment. Such real-time, goal-directed modulation of behavior is the key to acting as a mature adult. You need this capacity to free yourself from the automatic flow of experience, and to choose where to direct your attention. Without a director you are a mere automaton, driven by greed, fear, or habit.

Putting the Director Under the Microscope

The term some neuroscientists ascribe to the concept of the director is *mindfulness*. Originally an ancient Buddhist concept, mindfulness is used by scientists today to define the experience of paying close attention, to the present, in an open and accepting way. It's the idea of living "in the present," of being aware of experience as it occurs in real time, and accepting what you see. Daniel Siegel, one of the leading researchers and authors in this area, and co-director of the Mindful Awareness Research Center at UCLA, describes mindfulness as simply the opposite of mindlessness. "It's our ability to pause before we react," Siegel explains. "It gives us the space of mind in which we can consider various options and then choose the most appropriate ones."

To neuroscientists, mindfulness has little to do with spirituality, religion, or any particular type of meditation. It's a trait that everyone has to some degree, which can be developed in many ways. (It's also a state that you can activate, and that tends to become a trait the more you activate it.) Mindfulness also turns out to be important for workplace effectiveness. When you listen to a hunch that you need to stop emailing and think about how to plan your day better, you're being mindful. When you notice that you need to focus so you don't get lost driving to a meeting, you're being mindful. In each case you are noticing inner signals. The ability to notice these kinds of signals is a central platform for being more effective at work. Knowledge of your brain is one thing, but you also need to be aware of what your brain is doing at any moment for any knowledge to be useful.

Hundreds of scientists around the world now explore mindfulness, and one of the people central to this effort is Kirk Brown at Virginia Commonwealth University in Richmond, Virginia. As a graduate student, Brown noticed that some people were better than others at noticing internal body signals when recovering from medical challenges. A person who was aware of his internal experience seemed to heal from a tough operation faster than someone who wasn't. This awareness of signals coming from inside of you has a technical term: *interoception*. It's like perception of your internal world. Brown couldn't find an existing measure for this capacity to notice what was going on in your internal world, so he developed one, which he called the Mindful Awareness Attention Scale (MAAS). The MAAS is now the gold standard for measuring an individual's everyday mindfulness.

Brown discovered that everyone has the capacity for this type of awareness, but levels of mindfulness vary. As he tested people over the years, he found that people's MAAS scores correlated with their physical and mental health, and even with the quality of their relationships. "Initially we thought there was something wrong with our data," Brown explains. "It can't possibly be related to all these things. Yet all of the work we have done since supported this finding." Studies by Jon Kabat-Zinn, founding director of the Stress Reduction Clinic and the Center for Mindfulness in Medicine, Health Care, and Society at the University of Massachusetts Medical School, showed that people healing from skin diseases healed faster if they practiced mindfulness, and studies by Mark Williams at Oxford University found that the recurrence of depression could be decreased by 75 percent with mindfulness training. Mindfulness is clearly useful for getting and staying healthy, but is that just because it makes you less stressed, or is there something more powerful going on here? That's a question that Dr. Yi-Yuan Tang, one of the leading neuroscientists in China, wanted to answer. In 2007 he conducted a study to see if mindfulness was just a form of relaxation training, or if something else was at work. Forty volunteers underwent five days of mindfulness training for twenty minutes a day using a technique Tang calls integrative body-mind training. Another group did relaxation training for the same period. "There were significant differences between the two groups after only five days of training," Tang explains. The mindfulness group had almost 50 percent greater immune function on average, based on saliva samples. Cortisol levels were also lower in the mindfulness group. Mindfulness is clearly more than just relaxation. If so, what is it, and why does it have such a big impact on so many domains of life?

The Neuroscience of Mindfulness

A 2007 study called "Mindfulness meditation reveals distinct neural modes of self-reference," by Norman Farb at the University of Toronto, along with six other scientists, broke new ground in our understanding of mindfulness from a neuroscience perspective. To help you grasp the importance of this research, I'm going to recap first. You were born with the capacity to create internal representations of the outside world in your brain, called "maps." (These maps are sometimes called networks or circuits.) Maps develop based on what you pay attention to over time, such as Paul's map for credit cards. A lawyer would have maps for thousands of legal cases, a bushman from the Kalahari would have maps for finding water, and a young mother on her third child would have maps for how to get her children to go to sleep. We are also born with a strong capacity for certain maps to emerge automatically—such as the map for our sense of smell.

Farb and these six other scientists worked out a way to study how human beings experience their lives moment to moment. They discovered that people have two distinct ways of interacting with the world, using two different sets of maps. One set of maps involves the region mentioned earlier in the scene on distractions and insight. It's the "default network," which includes the medial prefrontal cortex, along with memory regions such as the hippocampus. This network is called default because it becomes active when not much else is happening, and you think about yourself. If you are sitting on the edge of a jetty in summer, a nice breeze blowing in your hair and a cold beer in your hand, instead of taking in the beautiful day you might find yourself thinking about what to cook for dinner tonight, and whether you will make a mess of the meal to the amusement of your partner. This is your default network in action. It's the network involved in planning, daydreaming, and ruminating.

This default network also becomes active when you think about yourself or other people; it holds together a "narrative." A narrative is a story line with characters interacting with one another over time. The brain holds vast stores of information about your own and other people's history. When the default network is active, you are thinking about your history and future and all the people you know, including yourself, and how this giant tapestry of information weaves together. In the Farb study, they like to call the default network the narrative circuitry. (I like the term *narrative circuit* for everyday use, as it's easier to remember and a bit more elegant than *default* when talking about mindfulness.)

When you experience the world using this narrative network, you take in information from the outside world, process it through a filter of what everything means, and add your interpretations. Sitting on the dock with your narrative circuit active, a cool breeze isn't a cool breeze, it's a sign that summer will be over soon, which starts you thinking about where to go skiing, and whether your ski suit needs a dry-clean. When the narrative network is active, we tend to easily leap from one idea to the next.

The default network is active for most of your waking moments and doesn't take much effort to operate. There's nothing wrong with this network; the point here is you don't want to limit yourself to experiencing the world only through this network.

The Farb study shows there is a whole other way of experiencing experience. Scientists call this type of experience *direct experience*. When the direct-experience network is active, several different brain regions become more active. This includes the insula, a region that relates to perceiving bodily sensations. Also activated is the anterior cingulate cortex, a region central to detecting errors and switching your attention. When this direct-experience network is activated, you are not thinking intently about the past or future, other people, or yourself, or considering much at all. Rather, you are experiencing information coming into your senses in real time. Sitting on the jetty, your attention is on the warmth of the sun on your skin, the cool breeze in your hair, and the cold beer in your hand.

A series of other studies has found that these two circuits, narrative and direct-experience, are inversely correlated. In other words, if you think about an upcoming meeting while you wash dishes, you are more likely to overlook a broken glass and cut your hand, because the brain map involved in visual perception is less active when the narrative map is activated. You don't see as much (or hear as much, or feel as much, or sense anything as much) when you are lost in thought. Sadly, even a beer doesn't taste as good in this state.

Fortunately, this scenario works both ways. When you focus your attention on incoming data, such as the feeling of the water on your hands while you wash up, it reduces activation of the narrative circuitry. This explains why, for example, if your narrative circuitry is going crazy worrying about an upcoming stressful event, it helps to take a deep breath and focus on the present moment. All your senses "come alive" at that moment.

Here's a quick exercise to try right now to make the research more meaningful. Find some incoming data to focus your attention on, just for ten seconds. If you are sitting down reading this book, focus on the feeling of sitting in your chair, paying close attention to the texture, springiness, and other aspects of your seat. Or focus on the sounds around you, observing the different sounds you can hear. Do this just for ten seconds right now.

If you did this exercise, perhaps you noticed several things, along with the incoming data you focused on. First, perhaps you noticed how hard it is to focus attention on one thing for ten seconds, which in itself is interesting. During the ten seconds, perhaps you lost track of the data you were trying to focus on and started thinking instead (which is the most common response to this exercise). At that moment, when your attention switched away from the feeling of the seat and went to your lunch, your brain switched from your direct experience to your narrative network. If you then remembered the exercise and went back to paying attention to your chosen data stream, you reactivated the direct-experience circuitry.

This quick experiment gives you a personal sensation of the shift between these two circuits, to be able to perceive the difference. If you did a similar exercise over and over, you would get better and better at noticing this shift as it happened. This occurs with people who practice types of mindfulness mediation. They get better at noticing the difference between directly experiencing something and the interpretation added by the brain. And doing these types of exercises regularly thickens the circuitry involved in observing internal states. Paying attention to a director makes him or her stronger and gives him or her more power.

Another thing you may have noticed in the ten-second exercise was other senses becoming more acute. When you sit on that jetty and stop to pay attention to the warmth of the sun on your skin, you soon notice the breeze, too. Activating the direct-experience network increases the richness of other incoming data, which helps you perceive more information around you. Noticing more information lets you see more options, which helps you make better choices, which makes you more effective at work.

Let's recap. You can experience the world through your narrative circuitry, which will be useful for planning, goal-setting, and strategizing. You can also experience the world more directly, which enables more sensory information to be perceived. Experiencing the world through the direct-experience network allows you to get closer to the reality of any event. You perceive more information about events occurring around you, as well as more accurate information about these events. Noticing more real-time information makes you more flexible in how you respond to the world. You also become less imprisoned by the past, your habits, expectations or assumptions, and more able to respond to events as they unfold.

Activating your director helps you perceive more sensory information. And here's where it gets more interesting. This sensory information includes information about your "self": information about your thoughts and feelings, emotions, and internal states. When you activate the director, you also notice more about what is going on inside you. And one of the most useful things to notice is what is happening within your own brain as you try to get work done: your stage being too tired to function, your stage getting too full, or your stage needing to switch off to allow an insight to get through. These types of observations become easier to perceive when you can activate your director at will.

The Point of Practice

In the Farb experiment, people who regularly practiced noticing the narrative and direct-experience paths, such as regular meditators, had stronger differentiation between the two paths. They knew which path they were on at any time, and could switch between them more easily. Whereas people who had not practiced noticing these paths were more likely to automatically take the narrative path.

A study by Kirk Brown found that people high on the mindfulness scale are more aware of their unconscious processes. Additionally, these people have more cognitive control, and a greater ability to shape what they do and what they say than do people lower on the mindfulness scale. If you're on the jetty in the breeze and you're someone with a strong director, you are more likely to notice that you're missing a lovely day worrying about tonight's dinner, and focus your attention on the warm sun instead. When you make this change in your attention, you change the functioning of your brain, and this can have a long-term impact on how your brain works. (The technical side to how that happens is something we'll get into in a later scene.)

Daniel Siegel explains it this way: "With the acquisition of a stabilized and refined focus on the mind itself, previously undifferentiated pathways of firing become detectable and then accessible to modification. It is in this way that we can use the focus of the mind to change the function and ultimately the structure of the brain." What Siegel is saying is that if you can activate your director at will, you perceive more information about your own mental state at any given time. You can then make choices to change what you pay attention to. And right here is the point of this intermission—and perhaps this book: By understanding your brain, you increase your capacity to change your brain. The more you notice your own experience, whether it's the small capacity of the stage, the dopamine high of novelty, or the way you need a moment to gather an insight, the more opportunities you have to become mindful, stop, and observe. Instead of becoming more self-aware by meditating on a mountain, you can do so while you work.

That's the good news.

Now for the bad news. Activating your director, as you will learn in the next act, is hard to do when there is a lot going on or when you feel under pressure. Some people go years without activating this circuitry, caught up as they are in the busyness of life. Activating your director at work isn't easy.

John Teasdale, recently retired, was one of the leading mindfulness researchers. Teasdale explains, "Mindfulness is a habit, it's something the more one does, the more likely one is to be in that mode with less and less effort . . . it's a skill that can be learned. It's accessing something we already have. Mindfulness isn't difficult. What's difficult is to remember to be mindful." I love this last statement. Mindfulness isn't difficult: the hard part is remembering to do it.

How do you remember to do something easily? You need to keep the director right at the front of the audience, so he can jump onstage fast when needed. It should be primed in your brain, something that's at the top of your mind because it was a recent experience. One of the best ways of having your director handy is practicing using your director regularly. A number of studies now show that people who practice activating their director do change the structure of their brain. They thicken specific regions of the cortex involved in cognitive control and switching attention. It doesn't matter so much what you use to practice. The key is to practice focusing your attention on a direct sense, and to do so often. It helps to use a rich stream of data. You can hold your attention to the feeling of your foot on the floor more easily than the feeling of your little toe on the floor: there's more data to tap into. You can practice activating your director while you are eating, walking, talking, doing just about anything, with the exception of drinking a beer in the sun, which works for only a limited time before your director leaves to go party. (The neuroscience of all that will have to wait for another book.)

Building your director doesn't mean you have to sit still and watch your breath. You can find a way that suits your lifestyle. My wife and I built a ten-second ritual into the evening meal with our kids that involves stopping and noticing three small breaths together before we eat. The added bonus is it makes a great dinner taste even better.

Having a director close to the stage helps keep your actors in line. As your director notices your brain's quirks in real time, you get better at putting words to experiences, which makes you faster at identifying subtle patterns as they occur. This skill increases your ability to make subtle changes. As your mind makes changes in brain functioning in real time, you become more adaptive, responding in the most helpful way to every challenge that comes along.

The lights are blinking; the bell is ringing; the intermission is over. Let's go back and watch the action on the stage as Emily and Paul face some new challenges. Let's find out how much a good director can improve a difficult scene.

ACT II

• • • • • • • •

Stay Cool
Under Pressure

The brain is much more than a logic-processing machine. Its purpose is to keep you alive. Every moment, your brain decides if the world around you is dangerous or helpful to this purpose. Sensing either danger or reward, even at surprisingly subtle levels, can have a dramatic impact on how and what you think. Automatic responses to dangers or rewards are thought of generally as emotions. Your ability to regulate your emotions instead of being at the mercy of them is central to being effective in a chaotic world.

In act 2, Paul discovers the impact of emotions on his thinking, then learns how to wrest back control when emotions take over. Emily learns about the brain's deep need to feel in control, and discovers a critical skill for managing stronger emotions. Finally, Paul sees that expectations play a role in how the brain processes information, sometimes having a big impact on how you perceive the world.

SCENE 7

• • • • • • • •

Derailed by Drama

t's 12:45 p.m. Paul hands the menus back to the waiter.

"So, do you think you can do it by then?" Miguel, the older executive asks. Paul is about to respond positively when the memory of an earlier project flashes into his mind. That client had asked for a tight time line, too. In the rush, Paul didn't find out what the client really needed, and ended up delivering late and over budget. Paul feels the frustration of that experience come to mind, as if he were right back there again. He doesn't want to let this emotion show. He tries to push down the sickening feeling in his stomach, but it doesn't seem to help. To make things worse, his narrative circuitry is aroused, and he is lost in internal thoughts, which is making him less aware of incoming information. He doesn't notice that a little too much time has passed since Miguel asked the question.

Paul takes another moment to think about what it would take to complete this project in eight weeks. He feels uncertain now, and he'd like to ask for twenty-four weeks. His emotions are all over the place, making it hard for him to think clearly.

"I think I can do it . . . ," he begins, "but is there any chance I could have, perhaps, just a little more time?"

Jill, the other executive, has a bemused look on her face. Her perfect nails and hair coiled in a bun remind Paul of a school head-mistress from his youth. He flashes back to a time when she held him in detention three days running, causing him to miss a school field trip. He wonders if Jill's look in response to his question was one of disdain. The heat rises in his suit.

"How well are you set up for this kind of project?" Jill asks.

Paul wishes he had turned off his phone and computer this morn-ing and focused more on preparing for this meeting, to be able to answer this question. Sweat beads on his forehead. He wonders if Jill will notice, which only makes him sweat more. He tries to make sure she can't see his discomfort, which takes focus, distracting his atten-tion from what she has just said.

"What was the question? Sorry," Paul replies, going a little red. "Oh yes. How we are set up. It's true, I'm a fairly small player," he re-plies. He can almost hear a quiet voice inside his head telling him he's worked on a comparable-sized project in the past, but he can't put a finger on which one. He hopes he might remember the project before the meeting is over.

"Look . . . I might not be the biggest company," he continues, "but at least I am local. This country will go down the tubes if we keep sending so much work offshore." As he finishes this sentence, he re-members a comment in the brief hinting at possible overseas competi-tors, but it's too late to retract his statement.

"Well, we love this country, too, but obviously, if we can get the project for a quarter the cost, we'd be crazy not to. It's the only way to compete against offshore retailers," Jill replies. Miguel nods.

The sinking feeling in Paul's stomach grows more intense. The meeting continues for another thirty minutes, with more tough ques-tions. Finally, Miguel and Jill thank Paul for his time. Paul smiles on the outside, but inside he is exhausted.

Getting back into his car, he unthinkingly takes the same com-plex route back home, only this time he gets lost. The intense banter from the meeting has exhausted his prefrontal cortex. He gets frus-trated trying to read a map while driving, and nearly hits a car slow-ing down for a yellow light. He gets home to find his son, Josh, sitting at the doorstep, home from school early. "What are you doing home so early?" he barks.

"Why wasn't your phone on?" Josh yells back at him. With all the intense emotions, Paul forgot he was supposed to be back early to meet Josh, home early from a school trip. Part of him knows he is in the wrong, but he can't help arguing with his son. "Don't you slam the door on me, young man," Paul yells. He wonders if he should start collecting a penalty each time Josh slams a door. An insight emerges out of nowhere, resolving a frustrating impasse from the meeting: a penalty, a fee, a toll. Of course. He had worked on a similar project before, darn it. The toll road project he worked on two years ago was almost the same as this credit card client, and it went well. If only he had remembered this during the meeting.

Paul is having a tough day. He is in a stressful situation that has gotten worse because of some quirks of the brain. He is experiencing emotional events from the past impacting performance in the present. Despite his attempts to wrestle his emotions to the ground, he failed to do so, and they affected his sales pitch.

Paul is operating under incorrect assumptions about how to regulate emotions. He thinks that trying *not* to feel something is the best strategy for *staying cool under pressure*. It's the "stiff upper lip" approach. He needs to change his brain's way of managing his emotions, so that he doesn't fall apart under pressure. Given that he wants to grow his business by doing more selling and less coding, this new set of circuits will be important to develop.

GOING LIMBIC

Human emotions are messy, involving many brain regions. Emotional experience is connected to a large brain network called the limbic system. The limbic system includes brain regions such as the amygdale, hippocampus, cingulate gyrus, orbital frontal cortex, and the insula, which are connected in various ways.

The limbic system tracks your emotional relationship to thoughts, objects, people, and events. It determines how you feel about the world, moment to moment. It drives your behavior, often quite unconsciously. Without a complete limbic system the human brain would be in poor shape, though basically functional. With no limbic system

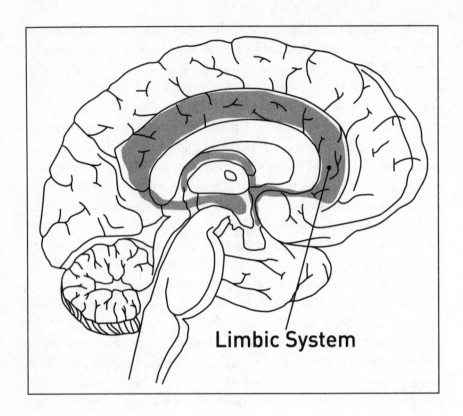

Limbic System

the basal ganglia could fire the right combination of motor neurons to get you out of bed, but once out of bed, you'd probably freeze. With a world of infinite choices each moment, there is not enough time or energy to logically process all the possible options for what to do next. Should you have breakfast? What should you eat for breakfast? Where should you eat? Or should you get more sleep? Back in bed, on the couch, or at the desk? Moment-to-moment decisions involve more than just rational processes. Subtle choices need to be made, based on value judgments. Executing these value judgments, such as deciding if a breakfast cereal is good or bad, is one of the limbic system's main functions.

TOWARD OR AWAY

Dr. Evian Gordon has developed one of the world's largest brain databases. Gordon has an intriguing vantage point as someone who can see patterns among a wide range of research. One of the central insights he has had, which he and Lea Williams put forward in their INTEGRATE *model*, is that the brain has an overarching organizing principle, which is to classify the world around you into things that will either hurt you or help you stay alive. "Everything you do in life is based on your brain's determination to minimize danger or maximize reward," Gordon explains. ' "Minimize danger, maximize reward' is the organizing principle of the brain."

The limbic system scans data streaming into the brain, telling you what to pay more attention to, and in what way. It's the limbic system's job to tell you whether some red berries on a bush are dangerous or tasty. Emotions such as curiosity, happiness, and contentment are *toward* responses. Anxiety, sadness, and fear, on the other hand, are *away* responses.

When the brain detects a threat that could endanger your life it is called a *primary threat*. Primary threats include real threats such as seeing a bear in the woods or getting hungry, hot, or thirsty, or even just seeing angry faces in a photograph. When your brain detects something that could help you survive, you experience a sense of reward, from noticing what are called *primary rewards*. Primary rewards include food, money, and sex, or even just a familiar face.

The limbic system is constantly making *toward* or *away* decisions—toward reward or away from threat. These decisions happen automatically, about half a second before you are consciously aware of them if you become aware of them at all. One study found that the brain does this even with nonsense words, which get classified as either positive or negative based on whether the phonemes, or sound units of the words, are perceived as pleasant or unpleasant.

As you experience emotions, your limbic system automatically becomes aroused. Many brain regions are part of this process, but the two more interesting ones are the hippocampus and the amygdale. The hippocampus is a large brain region involved in declarative memory, meaning memory that can be consciously experienced. Such memories are made up of billions of complex networks of neural maps, spread across the brain. The hippocampus is in charge of organizing and indexing these maps. Your hippocampus doesn't just remember facts; it also remembers feelings about facts. The stronger you feel about something, the easier it is to recall (with the exception of some events with intense emotions that are not remembered for more complex reasons). If you can remember far back enough to recall a mental picture of a favorite high-school teacher, you will also remember how you felt about her. The feeling arises at the moment the memory appears; it's all part of the same network.

The hippocampus is an important part of the network that remembers whether something is a danger or a reward, linking new experiences to previous memories. Paul had this type of limbic response when Jill reminded him of his own schoolteacher.

The amygdale is an almond-shaped region that sits just above the area responsible for smell. Although the amygdale is often thought of as the "emotional center" of the brain, it is just one part of the limbic system network. It works with the hippocampus and other limbic regions. The amygdale does have an interesting quirk that has helped make it famous: it tends to become aroused in proportion to the strength of an emotional response. It's like the brain's thermometer for feelings. And you can see this arousal clearly in fMRI studies. Arousal can be driven by either toward or away emotions, though, as you will see, these two types of emotions arouse the limbic system in different ways.

WALK TOWARD, RUN AWAY

In his book *The Happiness Hypothesis*, Jonathan Haidt writes that we are the descendants of people who paid a lot of attention when there was even a small rustle in the woods. In a dangerous world, it was the hyper vigilant who survived. Stimulate the amygdale with a probe (though I don't recommend you try this at home) and you mostly feel one group of emotions: *away* emotions such as anxiety. Though this anxiety could, of course, be a general response to having a probe in your brain, it's now accepted that the amygdale is somewhat neurotic, like a Woody Allen character: nervous, skittish, and easily thrown off kilter.

As well as being a lot more anxious than happy, the limbic system fires up far more intensely when it perceives a danger compared to when it senses a reward. The arousal from a danger also comes on faster, lasts longer, and is harder to budge. Even the strongest *toward* emotion, lust, is unlikely to make you run, whereas fear can do so in an instant. (Just put a plastic spider on someone's hand to observe this trait.) The *toward* emotions are more subtle, more easily displaced, and harder to build on, than the *away* emotions. This also explains why *upward spirals*, where positive emotions beget more positive emotions, are less common than downward spirals, where negative emotions beget more negative emotions. Human beings walk *toward*, but run *away*.

ISSUES, HOT BUTTONS, GREMLINS, HOT SPOTS, DEMONS

The limbic system gets aroused in a wide range of situations, some of which emerge later in this book. In this scene, Paul's limbic system has become aroused because it senses that a present situation is the same as a previous challenge. It's like walking along a stretch of path where a bear once jumped out at you. For Paul, the bear was a tight deadline, something that had bitten him, or at least bitten into his wallet, in the past.

Everyone has a unique set of "hot buttons" that can trigger limbic system arousal. These triggers have been discussed by psychologists and philosophers for centuries and go by many names, including the unconscious, patterns, gremlins, demons, and issues, but I'm going to call them *hot spots*. Hot spots are patterns of experience stored in your limbic system and tagged as dangerous. When the original pattern that produced the hot spot (or something similar) reappears, the danger response kicks in, proportional to the degree of danger tagged to the situation.

When overly aroused by real or imagined dangers (or the rarer strong rewards), the limbic system impairs your brain functioning in a number of significant ways. This reduced functioning often occurs without conscious awareness, and can even generate false confidence. For example, increased adrenaline when you experience fear might make you feel focused and therefore more confident in your decisions, when your ability to make the best decisions has actually been reduced.

THE IMPACT OF OVER-AROUSAL

When the limbic system gets overly aroused, it reduces the resources available for prefrontal cortex functions. If you could recall the name of a work colleague in one second without arousal, with arousal it might take five seconds, or you might not even be able to remember the name for an hour. The same happens for all the prefrontal cortex functions, including understanding, deciding, memorizing, and inhibiting. With less glucose and oxygen to get work done, the complex maps in your prefrontal cortex required for conscious processes don't function as they should. Your preexisting limitations get even worse.

The link between limbic system arousal and prefrontal function kicks in at surprisingly low levels. One study involved two groups of students completing the same paper maze, starting as a mouse in the middle of the page. One group had a picture of cheese, a reward, at the edge of the maze; the other, an owl. The groups then completed creativity tests. The group heading toward the cheese could solve around 50 percent more problems. Other studies show that prefrontal performance is impacted just from seeing smiling faces versus frowning faces at the end of sentences. It's really easy to set off the limbic system in a way that measurably reduces performance.

Paul's troubles began before he got to the meeting. When he arrived at lunch, he was already experiencing strong emotions, and didn't do anything to dampen them. His cognitive functions got worse when he remembered the project that went wrong before. As a result, he forgot an important point the client had mentioned in their first communication, that timing was central, and he mistakenly tried to ask for more time. Then he couldn't remember a previous similar project. This memory could have saved his presentation. It was only when he got home that his memory was triggered by a conversation with his son, Josh.

When there are not enough resources for conscious processing, the brain becomes more "automatic," drawing on either deeply embedded functions or ideas close to the front of your audience, such as recent events. Essentially your brain is just doing what it can with minimal resources, so it's using low-resource tools. For Paul, using a low-resource function resulted in his driving home the way he had driven to the meeting, because this was primed, at the front of his stage. This route was the wrong way to go, given that he was tired. He also forgot to turn his cell phone on.

Another challenge with increased limbic system arousal is that your director seems to go missing. Activating your director allows you to perceive more information and make better decisions. Good decisions are even more important when you are under pressure. Yet when the limbic system gets aroused, it becomes significantly harder to find your director. Ask someone in a meeting the question "Why do you think this way?" and they will generally have to pause and think hard to answer. It takes a lot of resources to think about thinking. It's like having four actors onstage, with four other actors noticing what the first actors are doing, and commenting on them. With room for only a few actors in total, or even fewer when the limbic system has reduced the stage's resources, this is tricky. Without his director, Paul finds it almost impossible to keep unwanted thoughts, such as the memory of the previous client, off the stage.

The third problem with limbic system arousal is that you become more likely to respond negatively to situations. You look at the downside, and you take fewer risks. The limbic system, superconscious of the dangerous side of life, looks out for even more danger when aroused by threats. As Paul experiences increasing arousal, it's more likely he will think the new project can't be done. He errs on the safe side with his estimates for the project, which, while helpful for project management, isn't the right frame of mind for selling his services. And in this negative state, it is also harder for Paul to have insights to help him solve impasses, such as how to answer a tough question about the capabilities of his company.

It's bad enough that an over-aroused limbic system gives you less space on the stage, and makes you more negative. But it gets worse. An aroused limbic system increases the chance of making links where there may not be any. In his aroused state, Paul finds himself thinking Jill looks like an old headmistress whom he disliked. When the amygdale is aroused it makes "accidental connections," misinterpreting incoming data. This misinterpretation happens through a rule of "generalizing." If you saw a snake recently, your brain becomes alert for objects even vaguely resembling snakes, including any kind of long, thin object. This happens because of the way the amygdale holds memories, which is at "low resolution," holding only a small amount of data. In the same way that it's faster to email a thumbnail of a photo than a large picture, by working at low resolution, the amygdale can respond to potential threats in milliseconds. This is a useful function when in danger. If you saw one snake, there's a chance there could be more, so it's best to stay on alert for anything even resembling a snake. But the amygdale's approximations of threatening memories also increase the likelihood of errors.

There's a second reason that accidental connections occur when you are anxious. There is a limitation to information processing called the *attentional blink*. This is the time gap required between identifying different stimuli. Most people's attentional blink is over half a second. You need half a second before the mind is free to think about something new. But if you hear a few words, and then your attention goes to an internal voice, as arousal tends to do, you might literally not have time to hear the next few words said to you. Dr. Craig Hassed teaches mindfulness training to medical students, because he finds not only does it reduce stress, but doctors who practice mindfulness make better decisions. "We literally don't see things when they are coming at us," Hassed explains. When you are anxious, you miss stimuli and make mistakes about what is said to you because your attention is going inward.

Here's one final blow to over-arousal. When you experience over-arousal over a long period of time, your allostatic load increases. This means your level of markers such as cortisol and adrenaline in the blood become chronically high. You experience a permanent sense of threat, and a low threshold for additional threats. Studies are showing that a high allostatic load can kill existing neurons, and stop the growth of new neurons in the hippocampus, important for forming memories. Clearly, being able to regulate your emotions well is not a "nice skill to have." It's essential for success, not just in work, but in life overall.

Fortunately there are brain-based techniques, tested and validated by neuroscience, that can reverse and even nullify the impact arousal. Just because a situation could over-arouse you, it doesn't have to be that way. There are several ways to minimize arousal, all of which involve the director intervening in the show in some way.

TIME IS OF THE ESSENCE

James Gross, associate professor of psychology at Stanford University, is at the forefront in the field of emotional regulation. Gross developed a model of emotions that distinguishes what happens both before an emotion arises and once it is present. He explains that before an emotion arises, there are several choices to be made: *situation selection, situation modification,* and *attention deployment.*

If Paul knew he was terrible at pitching to customers, he might have chosen not to pitch anymore and to hire someone else to do that task. That's situation selection at work. Once you're in a situation, you can modify it to some degree. That's situation modification. Paul could have chosen to do a sales pitch, but made sure he was thoroughly prepared. Even when you're already in a situation, you can still decide where to put your attention. That's attention deployment. Paul might have decided to do the pitch, and be prepared for it, but still have felt anxious, and chosen not to pay attention to this anxiety. This approach is similar to the way you manage distractions, the veto power I introduce earlier in the book.

These options work only before emotions kick in. Once emotions kick in, you have only three options. The first option is to *express* your emotions. If you're upset, cry, as kids do. Obviously, in many social and work settings, this doesn't work too well.

The second option is *expressive suppression,* which requires holding the feeling down and stopping the emotion from being perceived by others. Paul tried to suppress his emotions early in the meeting.

He was angry at himself for messing up with a previous client, and he tried not to let this show.

The third strategy involves *cognitive change.* "Even after you've got yourself into a bad situation, you can still, even at this relatively late stage, think about it differently," Gross explains. There are two examples of this phenomenon. One is called *labeling.* It's when you take a situation and put a label on your emotions. The other is called *reappraisal,* which involves changing your interpretation of an event. We will explore reappraisal in the next scene and focus here on labeling.

Gross set up lab experiments where people would watch emotion-inducing videos of scenes I won't bring to your mind right now. He would then get them to try different emotion-regulation techniques and evaluate the effects on the participants' emotional state, both by self-rated measures and by measuring bodily changes such as cortisol level and blood pressure. There are several surprising and important findings to this work. Gross found that people who tried to suppress a negative emotional experience failed to do so. While they thought they looked fine outwardly, inwardly their limbic system was just as aroused as without suppression, and in some cases, even more aroused. Kevin Ochsner, at Columbia, repeated these findings using an fMRI. Trying not to feel something doesn't work, and in some cases even backfires. Paul experienced this problem during his pitch, when he tried to suppress feeling bad about himself but ended up feeling even more anxious instead.

There's more. Gross found that when people try to suppress the expression of an emotion, their memory of events is impaired, as if they are consciously focusing their attention elsewhere, like watching television and paying attention to that while someone tries to talk to you. This happened to Paul. He lost the thread of the conversation and had to ask Jill to repeat her question. Trying to suppress the expression of an emotion takes a lot of cognitive resources, which leaves fewer resources for paying attention to the moment.

Gross had an observer sit across from the participants while they tried different emotion-regulation approaches. He found that when someone suppressed the expression of a negative emotion, the observer's blood pressure went up. The observer is expecting to see an emotion but gets nothing. This is odd, and in this way, suppression literally makes other people uncomfortable. "A bit like secondhand smoke, suppression has a real impact on others," Gross explains. Unfortunately, Paul is making the very people he wants to feel comfortable around him feel the opposite, because he doesn't understand how to regulate his emotions well.

So suppression has a lot of downsides, and expression is often out of the question. You can try to stay out of emotion-arousing events with situation selection, but that may lead to some downsides, such as not leaving the house much. The ability to veto where you focus your attention can help; however, there are times when you don't have the mental resources to do this, which is once an emotion kicks in. Sometimes you need to do more to wrestle an emotion down. What's needed is some form of *cognitive change*.

NAME THAT STATE

When your limbic system becomes aroused, the resources available for your prefrontal cortex decrease. However, this works the other way, too. Increasing the arousal of the prefrontal cortex can dampen down the arousal of your limbic system. The two work like a seesaw. You can make this switch happen by trying to find the right word to identify an emotional sensation, a technique that is called *symbolic labeling*.

Neuroscientist Matthew Lieberman, associate professor at UCLA, is another founding father of the social cognitive neuroscience field. He is also a leading expert on the link between limbic system and prefrontal cortex functioning, and he has done some groundbreaking research into labeling. In an important study in 2005, Lieberman and some colleagues asked thirty participants to view pictures of angry, scared, or happy-looking faces. Half the time the participants tried to match the target face to another picture of a face with a similar expression. The other half of the time, they tried to match the face to a word that correctly labeled the subject's emotion.

FMRI scans showed that when the participants labeled the emotional faces using words, less activity occurred in the amygdale. Interestingly, the part of the brain activated in this situation is the right ventrolateral prefrontal cortex, the region that is central to any type of braking in the brain, and the one that keeps reappearing as central to all types of inhibition. "This region goes on when you label," Lieberman explains, "and there is a correlated reduction in activity in the limbic system, including the amygdale, the cingulate, and the insula." The right ventrolateral prefrontal cortex becomes active even when people are not consciously trying to inhibit, as is the case in Lieberman's labeling experiment; all participants are doing in his experiment is stating what someone's face looks like.

Another study of labeling illustrates an intriguing quirk of human nature. Participants were asked to predict if they would feel better or worse if they spoke about their emotions. There was a strong tendency for people to expect that labeling emotions would result in increasing their emotional arousal. Surprisingly, people even predicted that labeling emotions would make the emotions worse, even after doing an experiment that illustrated that labeling their emotions decreased them! Because people incorrectly predict that voicing their feelings will make those feelings worse, a lot of people, especially in the business world, don't discuss their feelings. This is an example of humans developing some unfortunate habits from incorrect assumptions about human nature. We shouldn't be too hard on humanity, though. Plenty of studies show that speaking about emotional experience does bring the emotions back to the surface. The key is how you do it. To reduce arousal, you need to use just a few words to describe an emotion, and ideally use symbolic language, which means using indirect metaphors, metrics, and simplifications of your experience. This requires you to activate your prefrontal cortex, which reduces the arousal in the limbic system. Here's the bottom line: describe an emotion in just a word or two, and it helps reduce the emotion. Open up a dialogue about an emotion, though, and you tend to increase it.

David Creswell, a neuroscientist now at Carnegie Mellon University, also studies emotional regulation. He repeated Lieberman's labeling experiment. Only this time he first took measures of how mindful people were, using the Mindful Attention Awareness Scale. "In the people who are more mindful, we see an amygdale deactivation—it's actually turning off the amygdale completely," Creswell explains. He also found that with people who are more mindful, more of their brain becomes part of the inhibition process. "It wasn't just the right ventrolateral prefrontal cortex that got activated, but also the medial, right dorsolateral, the left ventrolateral prefrontal cortex (under the left temple), and other areas that got involved," Creswell says.

Being able to *stay cool under pressure* is a basic requirement for many jobs today. For people in leadership positions, this need is even more acute. Joan Fiore used to coach senior executives at Microsoft. "I try to imagine what it's like for these people to have to do what they do every day, and it just blows my mind," Fiore says. Most successful executives have developed an ability to be in a state of high limbic system arousal and still remain calm. Partly this is their ability to label emotional states. They are like an advanced driver who has a word for the experience of fear when he senses his car going into a skid. During a skid he can recall the word instantly, therefore reducing his panic. Stress is not necessarily a bad thing. It's how you deal with it that's key. Successful people learn to harness deep stress and turn it into *eustress*, thus enhancing prefrontal cortex functioning. They do this partly by *naming*, and using the other techniques coming up in the next scenes. People who succeed under pressure have learned to be in a place of high arousal but maintain a quiet mind, so that they can still think clearly. Over time and with practice, this capacity can become an automatic resource. The brain can be wired to deal better with emotions. Let's see the difference that better emotional regulation might have made to Paul during his pitch.

DERAILED BY DRAMA, TAKE TWO

It's 12:45 p.m. Paul hands the menus back to the waiter.

"So, do you think you can do it?" Miguel, the older client asks, looking at Paul.

"It's a heck of a time line," Paul responds, pausing to reflect for a moment. He flashes on a previous project that had gone wrong when that client was in a rush. As he notices his attention going there, he quickly inhibits this thought from taking up space on his stage, diverting his attention to the present clients and their facial expressions. Paul has a strong director, capable of observing his own thought processes in real time. He knows that a fraction of a second of focus on this past problem could generate runaway emotions, and that focusing on his senses can bring the narrative circuitry back in check.

With more attention at his disposal, Paul notices that a part of him wants to say it can't be done. At the same time, he wants the project. It would double his business. Yet he has no idea what it would take to complete the design and installation of the software in eight weeks. He'd like to ask for twenty-four. He takes a fraction of a second to step back and observe his thinking process and his emotional state, and finds he can assign a word to what's happening—he is feeling "pressure." Activating his director and putting a label on his experience has reduced the arousal in his brain. All of this happens in a less than a second.

With plenty of resources available for his prefrontal functions, Paul remembers that a team of developers in India were mentioned in the brief. Paul senses this means other suppliers are saying they'll do it in eight weeks. He weighs up two options, putting two groups of actors onto his stage to see which one he prefers. Actor one is to walk away. Actor two is to say yes now, and work out later how to do it. He compares the implications of both, picturing the outcomes each might bring. Because he is not in an overly stressed state, he remains optimistic, and just two seconds after his last comment, Paul blurts out, " . . . but I think I can do it."

Jill, the other client, has a bemused look on her face, but it doesn't bother Paul. He assumes it's her internal dialogue laughing at something, not at him. Her perfect nails and hair coiled in a bun remind Paul of a school headmistress in his youth, but he chuckles at this memory and lets the thought pass.

"How well are you set up for this kind of project?" Jill asks. He notices himself feeling defensive, but settles his emotions again by quietly recognizing and labeling his defensiveness. He can sense an idea brewing in the back of his mind and knows that he will need to be calm to remember the connection. In a flash he remembers a recent big project.

"Look, this isn't much bigger than one of my recent projects," he replies, slowing his breathing. "When I worked on the toll road to the east, two years ago, I built and installed the software to generate credit card payment from twenty thousand cars a day. We were on time and on budget. How many transactions a day would be happening over your network of stores?"

"About that number," Miguel responds, "but the difference is this would be across one hundred stores, not one location."

"Not a problem," Paul replies without missing a beat this time, keen to show his confidence. He leans forward. "Look, the technology for gathering the data together from five hundred locations is the easy part. Anyone can set that up. The devil is in the details, getting the software right within each store. I might not be the biggest company, but my strength is that I have done something similar before, so I can save you from the mistakes someone would make on their first attempt at this type of project. Plus, being small, I can work closely with your people, even come into your office every day if you like, to work through developments." Paul notices Jill making some notes about this point.

At the end of the meal, Paul is uncertain how the meeting has gone but is pleased with his performance. He knows he is tired, so he takes the main streets home so he doesn't have to think. A nice drive, operating on automatic, is what his stage needs to recharge. A few minutes later he remembers his cell phone is off, and he switches it on just in time to receive Josh's call to remind him about getting home early. At home, Paul and Josh play baseball for fifteen minutes, which helps Paul refresh his brain even more. Then Paul goes back to his desk to continue work on how he might deliver this job if he wins it.

Surprises About the Brain

- The brain has an overarching organizing principle to minimize danger (an *away* response) and maximize reward (a *toward* response).
- The limbic system can be aroused easily.
- The away response is stronger, faster, and longer lasting than the toward response.
- The away response can reduce cognitive resources, make it harder to think about your thinking, make you more defensive, and mistakenly class certain situations as threats.
- Once an emotion kicks in, trying to suppress it either doesn't work or makes it worse.
- Suppressing an emotion reduces your memory of events significantly.

- Suppressing an emotion makes other people feel uncomfortable.
- People incorrectly predict that labeling an emotion will make them feel worse.
- Labeling an emotion can reduce limbic system arousal.
- Labeling needs to be symbolic, not a long dialogue about an emotion, for it to reduce arousal.

Some Things to Try

- Use your director to observe your emotional state.
- Be conscious of things that may increase limbic system arousal and work out ways to reduce these, before the arousal kicks in.
- Practice noticing emotions as they arise, to get better at sensing their presence earlier.
- When you sense a strong emotion coming on, refocus your attention quickly on another stimulus, before the emotion takes over.
- Practice assigning words to emotional states to reduce arousal once it kicks in.

SCENE 8

• • • • • • • •

Drowning amid Uncertainty

t's 1:00 p.m. Emily has just finished lunch with Rick, the operations manager, and Carl, the company's finance director. The polite chatter about vacation plans is over and it's time for Emily to present her plan for a new conference. In her old job, she had a predefined budget and would execute a series of well-codified steps, involving getting sponsors, organizing speakers, and arranging marketing. In her new role, she creates these budgets and oversees others who run the conference. Her objective is to design three new "real world" conferences with parallel digital experiences, create budgets for them, and then beat these budgets. New conferences also have to be "sold" to other leaders in her organization. That's the purpose of this lunch.

Emily presents her first big idea, a conference about sustainability. She wants to bring business leaders together to discuss how to improve the long-term viability of companies in the face of economic challenges, climate change, and globalization. Though she is passionate about this topic, she's also anxious about getting the conference approved. There is so much uncertainty: whether the wider business world is ready for this idea, what they could charge attendees, who might speak, and who from her team would be the hands-on manager. She also feels uncertain about giving responsibility for the hands-on tasks to someone else after making sure all the details were handled herself for so long: Would anyone else do as good a job?

Women tend to be better at labeling their emotions. Emily knows she is anxious. Yet labeling alone hasn't settled down her limbic system. She still feels more tension than is helpful right now. Rick and Carl unconsciously sense this anxiety, which nudges their own limbic systems into alert mode. They start to question Emily's assumptions. As a result, her limbic system goes into overdrive. Now she is also uncertain why they are questioning her: Don't they trust her judgment? Is it because she is a woman? She senses her choices are being challenged, and she rails at the feeling of not having control of her work. She thinks back to a recent job where she was given a budget and left alone to oversee her domain.

The presentation of her next two projects doesn't go well. Emily tries hard to label and put aside her frustrations as they arise, but this tactic doesn't seem to be enough. She leaves the meeting wondering if this promotion is worth the pain.

Emily's challenge here is different from Paul's in the last scene. They both have to sell an idea, one of the most stressful parts of any job. Emily is more used to selling, so her limbic system has a lower baseline arousal level for this task than Paul's. He has spent more of his career behind a computer. In Paul's situation, his limbic system was overaroused by emotions from the past emerging in the present. Emily's limbic system is aroused by her anxiety about the future.

The brain craves certainty. A sense of uncertainty about the future and feeling out of control both generate strong limbic system responses. Emily is experiencing both of these threats at the same time during the lunch. For her to succeed in her new role, she needs to change her brain to recognize and deal with stronger emotions than labeling alone can handle.

THE ONLY CERTAINTY IS MORE UNCERTAINTY

Think of the brain as a prediction machine. Massive neuronal resources are devoted to predicting what will happen each moment. Jeff Hawkins, inventor of the Palm Pilot and more recently founder of a neuroscience institute, explains the brain's predilection for prediction in his book *On Intelligence*. He writes, "Your brain receives patterns from the outside world, stores them as memories, and makes predictions by combining what it has seen before and what is happening now.... Prediction is not just one of the things your brain does. It is the primary function of the neo-cortex, and the foundation of intelligence."

You don't just hear; you hear and predict what should come next. You don't just see; you predict what you should be seeing moment to moment. There is a popular email containing a paragraph of text with only the first and last letter of each word there, the rest junk. Yet most people can still read the contents. The brain is good at recognizing approximate patterns and making a best guess at what something means. This process of prediction happens with all the senses. It's how you can still hear people in a loud nightclub, for example. We "hear" even when we can't.

This predictive capacity, however, involves far more than just your five senses. Dr. Bruce Lipton, author of *The Biology of Belief*, says that there are around forty environmental cues you can consciously pay attention to at any one time. Subconsciously this number is more than two million. That's a huge amount of data that can be used for prediction. The brain likes to know what is going on by recognizing patterns in the world. It likes to feel certain.

Like an addiction to anything, when the craving for certainty is met, there is a sensation of reward. At low levels, for example, when predicting where your foot will land as you walk, the reward is often unnoticeable (except when your foot doesn't land the way you predicted, which equates with uncertainty). The pleasure of prediction is more acute when you listen to music based on repeating patterns. The ability to predict, and then obtain data that meets those predictions, generates an overall *toward* response. It's part of the reason that games such as solitaire, Sudoku, and crossword puzzles are enjoyable. They give you a little rush from creating more certainty in the world, in a safe way. Sometimes I call smartphones "dopamine delivery devices," in that they create this ability to get certainty on any topic quickly: the weather, the traffic, your stock portfolio, the state of the world. We feel more certain knowing we can get certainty in seconds about anything. Entire industries are devoted to resolving larger uncertainties: from shop-front palm readers, to the mythical "black boxes" that can supposedly predict stock trends and make investors millions. Some parts of accounting and consulting make their money by helping executives experience a perception of increasing certainty, through strategic planning and "forecasting." While the global financial crisis of 2008 showed once again that the future is inherently uncertain, the one thing that's certain is that people will always pay lots of money at least to *feel* less uncertain. That's because uncertainty feels, to the brain, like a threat to your life.

When you can't predict the outcome of a situation, an alert goes to the brain to pay more attention. An overall *away* response occurs. A 2005 study found that just a little ambiguity on its own lights up the amygdale. Think about someone you have spoken to a few times by phone, but never met or seen a picture of. You feel a mild uncertainty about him, yet even this tiny uncertainty seems to alter your interactions: notice how differently you interact once you know what that person looks like. Uncertainty is like an inability to create a complete map of a situation, and with parts missing, you're not as comfortable as when the map is complete.

Consider the uncertainty Emily is experiencing based on whether her sustainability conference proposal will be signed off on. The brain likes to think ahead and picture the future, mapping out how things will be, not just for each moment, but also for the longer term. Emily's brain tries to create two different futures: one where the proposal is signed off on and one where it isn't. Each map is enormous, and holding both in mind at once would be nearly impossible, as they involve similar networks. Emily would find herself switching between two giant maps, an exhausting process in itself. Also, not knowing if her project will be signed off on would feel to Emily as if something were stuck in her decision queue. Once this decision is made, many other decisions that her brain wants to make will be easier.

For Emily, the uncertainty of not knowing if she will sell her conference idea, not knowing where and when the conference will be, and not knowing who will run it has reduced her capacity to be her best. Her colleagues notice this. She needs stronger emotional regulation techniques to manage uncertainty. However, before getting to these techniques, let's explore one other factor Emily experienced that made matters worse.

AUTONOMY AND THE PERCEPTION OF CONTROL

As well as anxiety from a sense of uncertainty, Emily also experienced stress from recognizing that her level of control over her work had decreased. She now has to get multiple people to sign off multiple times, and she has to get others to do her work, rather than run the conference herself. Though she is in a more senior role, her perception of autonomy, of being able to make her own choices, has decreased.

Autonomy is similar to certainty, and the two are linked. When you sense a lack of control, you experience a lack of "agency," an inability to influence outcomes. A sense of not being able to determine the future, to predict what will happen moment to moment, emerges. This feeling, of course, generates more uncertainty. However, certainty and autonomy also appear to be individual issues. You can be stressed by a lack of certainty but still have a lot of autonomy, like Paul, who is his own boss but can't predict his revenues until he closes deals. Or you can have a lot of certainty from a secure job, but a micromanaging boss may not let you make decisions, giving you a low sense of autonomy. Good apps on a smartphone do both, giving you information—which is certainty—with the least effort possible, and then autonomy in the form of having simple choices to make, again as effortlessly as needed. When you access Waze or other navigation apps, for example, you get to see traffic issues in real time and are given options for what to do now.

A sense of autonomy is a big driver of reward or threat. Steve Maier at the University of Boulder, in Colorado, says that the degree of control that organisms can exert over something that creates stress determines whether the stressor alters the organism's functioning. His findings indicate that only uncontrollable stressors cause deleterious effects. Inescapable or uncontrollable stress can be destructive, whereas the same stress that feels escapable is less destructive, significantly so. Steven Dworkin, a professor in psychology at the University of North Carolina at Wilmington, studies the way rats are affected by drugs. In one study, a rat gives itself cocaine directly by pressing a lever. The rat dies from lack of food and sleep. The surprising part is what happens when a second rat gets the same dose of cocaine at the same time as the first rat, but not of its own volition. It dies sooner. The difference is a perception of control (or so scientists think; the rats don't say much.) Jokes aside, this type of study has been done with electric shocks and other stressors, and even on humans (not to the point of death, of course). Over and over, scientists see that the perception of control over a stressor alters the stressor's impact.

And there's more. A study of British civil servants found that low-level, nonsmoking employees had more health problems than senior executives. This doesn't make sense intuitively, as senior executives are known to be under a lot of stress. It appears that the perception of choice may be more important than diet and other factors for health. Choosing in some way to experience stress is less stressful than experiencing stress without a sense of choice or control.

A number of studies show "work-life balance" as the main reason people start their own small businesses. Yet small business owners often work more hours, for less money, than in corporate life. The difference? You are able to make more of your own choices, or at least it feels like that. Yet another study, looking at residents in a retirement home, found the number of deaths was halved in the study group, compared to a control group, when participants were given three additional choices about their environment. The control group were people on a different floor on the same premises. The choices themselves were not significant: a different plant or a different type of entertainment, for example. There are even now long-term global studies of the impact of autonomy on well-being. One study showed that psychological prosperity (such as a sense of autonomy) generates feelings of well-being more than economic prosperity. Another study found that employees with a greater sense of autonomy had higher job satisfaction and overall reduced stress.

Amy Arnsten at Yale Medical School studies the effects of limbic system arousal on prefrontal cortex functioning. She summarized the importance of a sense of control for the brain during an interview I filmed with her at her lab at Yale. "The loss of prefrontal function only occurs when we feel out of control. It's the prefrontal cortex itself that is determining if we are in control or not. Even if we have the illusion that we are in control, our cognitive functions are preserved." This perception of being in control is a major driver of behavior. In fact, in study after study, we see that a feeling of control can literally be a life or death issue.

MAKING CHOICES

Another way of thinking about autonomy is through the lens of being able to make choices. When you sense you have choices, something that used to feel stressful now feels more manageable. Finding that you have choices in a situation reduces the threats from both autonomy and uncertainty. Emily could reduce her stress over her proposal being approved by remembering she could choose to reschedule her meeting, and is choosing to present her ideas today. Even the smallest perception of choice seems to impact limbic system arousal. Imagine you are frustrated by your boss telling you that you have to hire a new team member and feel you have no choice even though it will take too much of your time. If you take a quiet moment and find a positive reason to hire someone (to reduce your long-term workload, for example), your limbic system will shift to more of a *toward* response. In this toward state, it's much easier to reflect on your situation.

This idea of the importance of a perception of choice is easy to test on children, who often rail against a lack of choice. When a child won't go to bed, you might reduce her resistance by giving her back a choice. For example, she can choose whether she is read a book or told a story. This choice can have a big impact. It's the "perception" of choice that matters to the brain. Studies of teenage behavior shows that the terrible teens is not a biological necessity, as a number of cultures don't experience this phenomenon. A study of teenagers in Western cultures found that these teenagers have fewer choices than a felon in prison. Food for thought.

Finding a way to make a choice, however small, seems to have a measurable impact on the brain, shifting you from an *away* response to a *toward* response. If this seems strange, know that the act of pushing an object away versus pulling the same object toward you also creates this kind of change in the brain. Emotional states are surprisingly easy to change at times. One different word or phrase can make a big difference.

If I am driving in traffic and allow myself to get annoyed at being delayed, in this brain state, small frustrations, such as forgetting a document, become bigger. At some point my director might kick in (perhaps I look in the mirror and notice how grumpy I look). I might decide to let go of my frustration and focus on relaxing while I drive, knowing I want to write later that night, which won't happen if I am exhausted from feeing grumpy. I decide to be responsible for my mental state instead of a victim of circumstances. In the instant that I make this decision, I start seeing more information around me, and I can perceive opportunities for feeling happier, such as remembering to call a friend. This experience is one of finding a choice and making that choice, and it shifts what and how I perceive in that moment.

Much has been written about the importance of "taking responsibility" in life. *Responsibility* means an ability to respond. Generating a *toward* response, by making an active choice, increases your ability to respond to incoming data in adaptive ways. This concept is important for maximizing performance at work, as there are so many situations that can over-arouse the limbic system. This idea of consciously choosing to see a situation differently is called "reappraisal," and it's the missing link for Emily at her lunch meeting.

EVERY DARK CLOUD HAS A SLIVER OF REAPPRAISAL

Cognitive reappraisal (reappraisal, for short) is the other cognitive change strategy for regulating emotions. A series of studies shows that reappraisal generally has a stronger emotional braking effect than labeling, thus it's a tool for reducing the impact of bigger emotional hits.

Reappraisal often goes by other names, such as reframing or recontextualizing. There are all kinds of aphorisms for reappraisal, such as turning a sow's ear into a silk purse or finding the silver lining in a dark cloud. Kevin Ochsner, at Columbia University, studies the neuroscience of reappraisal, building in part on James Gross's psychological research. "There's a famous finding in the psychological literature," Ochsner explains, "showing that six months later, someone who has become a paraplegic is just as happy as someone who's won the lottery. It seems clear people are doing something to find what's positive in even the most dire of circumstances. The one thing you can always do is control your interpretation of the meaning of the situation, and that's fundamentally what reappraisal is all about."

In one of Ochsner's reappraisal experiments, participants are shown a photo of people crying outside a church, which naturally makes participants feel sad. They are then asked to imagine the scene is a wedding, that people are crying tears of joy. At the moment that participants change their appraisal of the event, their emotional response changes, and Ochsner is there to capture what is going on in their brain using an fMRI. As Ochsner explains, "Our emotional responses ultimately flow out of our appraisals of the world, and if we can shift those appraisals, we shift our emotional responses." While most reappraisal tends to be toward being more upbeat, it's also possible to reappraise negatively, to alter a perspective for the worse. Emily did this at lunch, deciding that her colleagues' questions meant they were questioning her judgment. Remember that perceived dangers pack a punch, so even a small reappraisal in the wrong direction can have quite an impact.

Ochsner's research finds that as people reappraise positively, there is increased activation of the right and left ventrolateral prefrontal cortex, and a corresponding reduction in activation of the limbic system. This is similar to what Lieberman finds when people label emotions. It turns out that conscious control over the limbic system is possible, not by suppressing a feeling, but rather by changing the interpretation that creates the feeling in the first place. One difference between labeling and reappraisal, though, is that while people incorrectly predict that labeling will increase arousal, they correctly predict that reappraisal can reduce arousal.

A REAPPRAISAL FOR ALL SEASONS

From my own observations I believe there are four main types of reappraisal. The first type is what happens in the wedding/funeral picture experiment. You decide that a threatening event is no longer a threat. We do this type of reappraisal a lot, usually without knowing it. For example, when I am at an airport walking toward a gate I can't see, I get anxious about missing my flight. Once the gate is in sight and I can see a queue of people, my anxiety drops. I have decided I am not in danger, and I immediately feel better. This first type of reappraisal involves *reinterpreting* an event.

I am reminded of Philippe Petit, the tightrope walker who walked between the World Trade Center's twin towers in the 1970s. He worked out a way to manage his fear of the height by hiring a helicopter and spending time sitting in its open doorway high above the towers, getting comfortable with being even higher than he was planning to walk. Making his brain decide that it wasn't dangerous a thousand feet above the towers made it possible to feel "safe" on the wire below a few days later. The wire no longer felt so high! Think of this type of reappraisal as changing your raw emotional response to an event.

The second type of reappraisal is at the heart of many effective management and therapeutic techniques. It goes by the name of *normalizing*, and it's a widely useful tool. Let's say you are in a brand-new job and don't yet have mental maps for even simple things such as finding stationery or coffee. Everything is new. New means uncertain, which means arousing, which reduces space on your stage. Yet being in a new environment means you need to use your stage a lot. With overworked actors, your ability to label or reappraise, to dampen down the arousal of uncertainty, is harder. In this way, doing anything significantly new can create a negative spiral. This is one of the reasons why change is so hard: doing things differently can bring about a negative spiral that can feel overwhelming.

If Emily had known that it was "normal" to feel overwhelmed in the first few weeks of a new job, her sense of uncertainty would have decreased. Having an explanation for an experience reduces uncertainty and increases a perception of control. The field of change management builds on the power of normalizing, by describing the emotions and stages that occur during change, such as denial or anger, to help people reduce the threat response. When you normalize a situation, be it the stress of a new job or the challenges of bringing up teenagers, you are using a second type of reappraisal.

The third type of reappraisal is a little more complex, but essentially it involves *reordering* information. The brain keeps information in nested hierarchies. All information is positioned relative to other ideas. This is similar in a sense to how an organizational chart looks: every map in your brain is above some maps and below or alongside others. For example, Emily has valued the map for "family" as being more important than her map for "work." As it turns out, she also values being left alone to do her job more than the idea of working with others.

This new job is challenging Emily's ordering of things. She wants to run the sustainability conference, but to do so she needs to work more with others. Yet she values working alone more. Something has to give. By looking afresh at the values she ascribes to situations, Emily might find a way to put a positive spin on working with others, to raise the value of this idea. This kind of reappraisal would result in massive numbers of neurons in her brain being reordered into new hierarchies, in relation to massive numbers of other neurons. This cognitive change tends to come with a big release of energy, perhaps due to the amount of reconfiguration going on. Reordering how you value the world changes the hierarchical structure of how your brain stores information, which changes how your brain interacts with the world.

The last type of reappraisal is probably the hardest to do, but at times can be the most effective. It is similar to reordering, though appears to require more space on the stage to be applied. It's so easy to become fixed in a way of thinking, as you discovered in the scene on impasses and insights. One of the most common causes of tension between people is someone being fixed in his own worldview and not being able to see the world through another person's eyes. When you take another person's perspective, you are changing the context through which you view a situation. This is something Emily could have done at her meeting, perhaps seeing herself through the eyes of her colleagues, who didn't yet know much about her, instead of assuming they didn't trust her. Think of this type of reappraisal as repositioning, as you are finding a new position from which to look at an event. It could be from another person's position, or from another country or culture's perspective, or even from a perspective of yourself at another time.

Each of these four types of reappraisal—reinterpreting, normalizing, reordering, and repositioning—are techniques people use all the time. With a deeper understanding of the biology behind reappraisal, and thus richer, easier-to-find maps for these techniques, you can begin to reappraise more often, and more quickly, which can significantly increase your ability to stay cool under pressure.

REAPPRAISAL AS THE "KILLER APPLICATION" FOR EMOTIONAL REGULATION

In act 1, I introduce the idea of being at the top of the inverted *U*, that optimal level of arousal for making decisions and solving problems. This is a state of quiet alertness, one in which you are able to think on several levels at once. If there is space for the director to jump in from time to time and observe a mental process as it's occurring, your thinking will improve even more.

Okay, so that's the "perfect world" that none of us lives in. Work involves complex, uncertain, messy tasks. Someone who could not regulate his emotions well would last about an hour in most jobs. Yet while most people have reasonable emotion-regulation capacities, they still operate under more arousal than is ideal for peak performance. With too much arousal, the director is hard to find. Without the director, the mind easily wanders, and it's easier for irrelevant audience members to jump onstage and take over. A small amount of overarousal can result in your taking longer to do simple work or missing important insights. This is particularly an issue in an era when we are bombarded with potentially threatening information through our smartphones, from news updates to looming social conflicts with family members on Facebook to a constant flow of information from work via email. This always-on situation, where any spare moment is taken up by checking your phone, makes our brains noisier and more prone to be already in a threat state when a possible threat comes along. Being always on primes us for downward spirals.

It doesn't have to be this way. As you learn more about your brain, it becomes possible to stay calm in just about any situation, including the overwhelming limbic system arousal driven by uncertainty about the future and the constant barrage of information coming from technology (though you may need to switch off more often as well, a type of situation modification). It is reappraisal that gives you this capacity.

Consider what Kevin Ochsner said when I asked him about the impact of his reappraisal research on his own thinking. "If our emotional responses fundamentally flow out of interpretations, or appraisals, of the world, and we can change those appraisals, then we have to try and do so. And to not do so, at some level, is rather irresponsible."

Let's explore how central reappraisal is to success at work with Emily's situation. She is uncertain if she will sell her sustainability conference idea to her colleagues, and this thought is making her anxious and ineffective. Going back to Gross's list of emotion-regulation options, she could try situation selection, by sending someone else to sell the conference idea, but that might not go down so well. She could do situation modification, perhaps by holding the meeting out in the

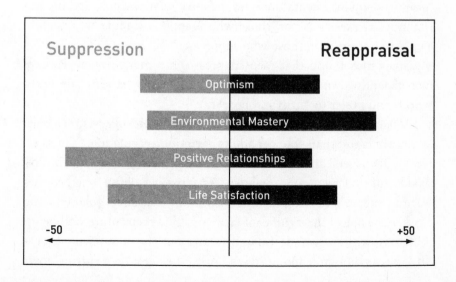

sun in a park, but she might still get anxious there. She could try shift-ing her attention by not focusing on her anxiety, but her arousal might be too strong for that. She could try to express her emotions, but you can imagine how well that would go down. She could try suppressing her feelings, but she would still be anxious, perhaps even more so, and her colleagues would feel anxious, too. The best options for Emily involve cognitive change. Labeling her emotions doesn't seem to be enough. Which leaves her with reappraisal.

Emily might reappraise by noticing she feels anxious about selling the idea to her colleagues, and deciding that she doesn't need to sell to them at all. She decides to ask for their help instead. Or she might decide to consider her colleagues as people who will notice things she hasn't been able to see for herself, so that when she presents to the CEO she will have covered all the angles. Changing her interpretation of the event in one of these ways will probably change the outcome of the meeting, resulting in the holding of an important conference that wouldn't happen otherwise. Perhaps Ochsner is right: sometimes it's irresponsible not to practice reappraisal.

The research on reappraisal shows that it's a strategy with few if any downsides, and significant upsides. Gross did another study out-side the lab where he grouped hundreds of people based on whether they tend to use reappraisal or suppression to deal with emotions. He then compared these two groups on a wide range of tests, including optimism, environmental mastery, positive relationships, and life sat-isfaction. On every factor, those who reappraised more were signifi-cantly better off than those who suppressed.

Gross also found that men suppress more than women. Perhaps men generally think it's not "manly" to tell oneself "stories" about the world, and prefer to "grin and bear it."

"There is a substantial body of research which suggests that older adults are better than younger adults at emotion regulation," Gross ex-plains. Teenagers' ability to express their emotions is one of their most wonderful and terrifying qualities. As teens get older, whether they learn to suppress or reappraise as their main emotion-regulation strat-egy may be one of the significant factors behind their future well-being.

Gross, with the wonderful understatement of a pure scientist, says, "It looks as though reappraisal is a fairly efficient way of cutting down the experience and biological representation of negative emotion." He may be too subtle. To me, reappraisal is one of the most important skills needed for success in life, the other being the ability to observe your mental processes. When I asked Gross what he thought about reappraisal and its role in education and wider society, he was more effusive: "I think this knowledge should be taught early, and often. It should be in the water we drink."

Yet, while it might sound like reappraisal is the way to bring about world peace and end hunger, this technique also poses some challenging philosophical questions. In 2007, I presented the research on reappraisal to doctors at a cancer research institute. "Are you trying to say," a senior scientist challenged me, "that success at work is based on your ability to make up false interpretations of the world instead of dealing with reality?" I had to pause and reflect for several moments before answering. Research shows that people who see life through slightly rose-colored glasses do in fact seem to be the happiest. And happy people perform better at many types of work. The answer to the doctors' question is, basically, yes, (though, of course one can take this too far). To a logical, fact-based scientist, this answer can be difficult to accept. Reappraisal requires cognitive flexibility, the capacity to see things from many angles, which more creative types tend to be good at. To a technician, the idea of creatively seeing other perspectives is not just illogical but also perhaps just a little foreign, and therefore a little uncertain.

However, there is another angle to this, a reappraisal if you like. Consider this quote from one of the great neuroscientists, Walter Freeman: "All the brain can know it knows from inside itself." If you recognize that all interpretations of the world are only that—interpretations your brain has made, and ultimately just yours—then having a choice about which interpretation you might use at any moment makes more sense.

There is some bad news about reappraisal, which partly explains why it's not everyone's tool of choice. Reappraisal is metabolically expensive. It's not easy to do, especially if your stage is full or your actors are tired. With reappraisal, first you have to inhibit your current way of thinking, which requires a lot of resources. Next you have to generate several alternative ways of thinking, each a complex map, and hold these alternatives in mind long enough to make a decision among them. Then you have to choose the alternative interpretation of events that makes the most sense, and keep focused on that. This all points to the need for a strong director. Without the capacity to use your full cognitive power at will, your capacity to reappraise will be limited to moments when you are well rested.

The effort involved in reappraisal explains why it tends to be easier to reappraise with someone else. Many of the tools and techniques in mentoring, coaching, career counseling, or various therapies attempt to change your interpretation of events. Another person sees things about you that you can't. It's like having a bonus prefrontal cortex.

The other way to make reappraisal easier is through practice. The more you practice reappraisal, the less effort it takes, as you develop thicker networks between the prefrontal cortex and the limbic system. Coaching helps people practice reappraisal. Optimists may be people who have embedded an automatic positive reappraisal to life's knocks. Optimists dampen their over-arousal before it kicks in, always looking at the bright side before a nagging doubt takes over.

Humor may also be a form of reappraisal. John Case, a retired CEO I know, had a phrase he used when people got tense in a meeting: "Did I tell you I just got a great deal on car insurance?" Out of context, this comment made people crack up, which shifted their perspective from serious to funny. From away, to toward. You have probably noticed how much easier it is to see options when you laugh at an otherwise tough situation. With humor, you don't need that cognitively expensive step within reappraisal of trying to scroll through lots of alternative perspectives and come up with the perfect new perspective that triangulates all your different objectives: just choose the perspective that makes you chuckle. In this way I like to think of humor as a type of cheap reappraisal.

REAPPRAISING YOUR OWN BRAIN

Let's take this to another level. Frustration at your limitations, mistakes, missed opportunities, forgetfulness, or bad habits can generate a lot of limbic activity. A common automatic response when people get annoyed with themselves is to try to suppress this feeling, to push aside the internal frustration. But you now know what suppression can do for emotions.

This brings us to a central idea for this book. As you learn more about your brain, you begin to see that many of your foibles and mistakes come down to the way your brain is built. You can't think about a complex work situation and walk around the house at the same time (which I discovered the hard way, jamming my toes under the door.) It's not you; it's your brain. You can't learn to do anything new and complex—such as learn to ride the subway in Japan without an interpreter—without your limbic system firing up from uncertainty, and in this state you're going to make mistakes (which I also learned the hard way, getting rather lost one day). It's not you; it's your brain. And you can't go into a meeting at four o'clock in the afternoon and expect yourself and everyone else to come up with brilliant ideas. It's not you or them; it's your brain.

So, next time you're tough on yourself, you can say, "Oh, that's just my brain." This statement is an act of reappraisal in itself. This approach may be a far better strategy than trying to express your emotions. And it's probably much better than trying to suppress your frustration about your own foibles. Like using humor to reappraise, this strategy is a quick and easy one to remember and to activate, which is important when the stage's resources are being taxed.

With all this in mind, let's see how Emily's lunch might have gone if she had noticed she was feeling uncertain and out of control, and had found a way to reappraise to reduce her strong arousal.

DROWNING AMID UNCERTAINTY, TAKE TWO

It's 1:00 p.m. and lunch has just finished.

Emily presents her first big idea, a conference on sustainability. She wants to bring business leaders together to discuss how to improve long-term viability of companies in the face of climate change and globalization. Though she is passionate about this topic, she's also anxious about getting the conference approved. There is so much uncertainty: whether the wider business world is ready for this idea, what they could charge attendees, whom they might get to speak, and who from her team would be the hands-on manager. She also feels uncertain about turning over responsibility for the hands-on tasks to someone else after having done them for so long: Would anyone else do as good a job?

Emily notices that all this uncertainty is increasing her anxiety. This noticing is an act of labeling her emotions, which helps a little. Next she tries to veto her attention focusing on her anxiety, but this doesn't seem to shake her state. She has to find another way to view her situation. She reflects for a moment and sees a few ways she could view this meeting, and settles on one idea: that it's a chance to get to know her new bosses and see how she might best work with them. She has reinterpreted the situation, and with this type of reappraisal, her limbic system dampens down.

Emily notices Rick and Carl questioning her assumptions, and is about to get defensive when she decides to veto the defensive feeling, which she is able to do in this calmer state. She reappraises the situation, this time looking at herself from Rick and Carl's perspective, a type of repositioning. From this point of view, she can see it would be important that her bosses look closely before investing serious company funds, especially with someone new creating the budget. They will probably cut her slack after she proves herself. With this in mind, she doesn't react defensively to their questions, and the questions stop after a few minutes. She presents all three conferences well, and feels pleased with her performance. By the end of the hour they have agreed in principle to the sustainability conference and decided on a date for the event. She is ready to present the idea to her team and to choose who might be the best person for the job.

Surprises About the Brain

- Certainty is a primary reward or threat for the brain.
- Autonomy, the feeling of control, is another primary reward or threat for the brain.
- Strong emotions generated by certainty and autonomy may need more than labeling to be managed.
- Reappraisal is a powerful strategy for managing increased arousal.
- People who reappraise more appear to live better lives.

Some Things to Try

- Watch for uncertainty creating a feeling of threat; practice noticing this.
- Watch for a feeling of reduced autonomy creating a sense of threat; practice noticing this.
- Find ways to create choice and a perception of autonomy wherever you can.
- Practice reappraisal early when you feel a strong emotion coming on.
- You can reappraise by reinterpreting an event, or reordering your values, or normalizing an experience, or repositioning your perspective.
- Reappraising your own experience is a powerful way of managing internal stressors; use this technique when you are anxious about your mental performance by saying, "That's just my brain."

SCENE 9

• • • • • • • •

When Expectations Get
Out of Control

I t's 3:00 p.m. Paul is back at his desk, trying to plan how he will deliver the new project if he wins it. He agreed to the tight deadline, and asked for two days to create a detailed project plan before providing a final formal quote. Since the client first contacted him four days ago, Paul has been looking forward to working out how much money he could make from this project. He expects a healthy profit, enough to pay for a good vacation and to take his business to another level. The potential windfall has helped generate a positive mood since Paul first thought about it, and he has been chatting with Emily about where they might take a vacation this year with the extra money. He's also pleased he told his contractors about the pitch. He hasn't had much work for them lately, and the news seems to have cheered them up.

Paul opens up a spreadsheet to build a proper budget for the project. He puts in the highest fee he thinks he can charge and still be competitive, and works out his general costs. After a few more calculations, he notices he will need to use all his contractors if he is going to write the software in eight weeks. As he gets close to finishing, he is looking forward to scrolling down the spreadsheet to see how much profit there will be. Finishing his calculations ten minutes later, he scrolls down and sees that the profit number is in the negative. Paul isn't too worried at first. He assumes there must be an incorrect formula somewhere, and he goes looking for his error.

Twenty minutes later, he stares into the kitchen sink, the tap on, watching the water run. He has been standing like this now for more than two minutes.

"Dad, there's a drought on, you know," Josh yells from the open fridge door, where he is hunting for a snack.

"Oh yeah," Paul replies distractedly.

"I'm going down to the store. There's nothing here to eat. Can I have some money, please?" Josh asks as he shuts the fridge door.

"No, go do your homework," Paul replies. "And there's plenty of food. We spent a fortune on shopping two days ago."

"Dad, what's your problem? You are usually happy to get me out of the house. Don't be such a pain."

"Look," Paul responds, now getting uptight, "just do what I say. It's not a good day for me."

"But, Dad, I've arranged to meet a friend."

"Well, tell them your dad is a nasty person and won't let you go."

"Fine." Josh storms off. Moments later his bedroom door slams.

Paul goes back to his office and tries to think. He can't put the price up, so he will either have to let the project go or find a cheaper way of delivering it than though his usual suppliers. Neither option makes any sense right now. A wave of depression washes over him. Irrelevant tasks soon distract him, things his assistant should be doing: opening envelopes and filing papers. He wants to do something to take his mind off feeling so glum, so he starts writing a note to his suppliers about the job not going ahead for now. As he writes, a quiet alert signal about emailing such a note almost makes it into his awareness. But, like a cell phone at a loud party, the signal is too quiet to be noticed. He presses Send.

Moments later, a reply arrives from a long-term supplier, Ned. Ned claims Paul is being blinded by the money. Paul shoots an angry email back.

Thirty minutes later, while Paul is replying to another angry email from a different supplier, Michelle gets back from school. She asks her father about his day, and they talk about what happened. Michelle is only three years older than Josh but seems a decade more mature.

"Dad, why don't you just find someone overseas to do the coding for you? That's what people do nowadays," she suggests.

"Thanks for the thought, sweetie, but I don't know anyone I could trust. Plus I'd have to go overseas myself, and there's no time to do any of that."

"Maybe there's another way," Michelle says as she wanders off to the kitchen. She finds ingredients for a sandwich and makes them both a snack.

They go to the back porch to eat. Paul asks Michelle more about her day. Michelle got better marks on an art project than she had expected. Paul gets interested in what she's doing, wondering whether she might be talented. For a moment he pictures himself back at school working on a science project, excited about learning. An idea bursts onto his stage: he might be able to find a supplier who arranges software programming for small consultants like him. He goes back to his desk to search for suppliers online. Within a short time he has sent inquiries to three companies that appear reputable, and already has one response back. He starts to feels better already. The fog of depression lifts, leaving a slight haze of possible good news in its stead. If only he hadn't made such a mess of things.

In under an hour, Paul has managed to harm some important relationships, with both his son and his long-term suppliers. He'll be able to mend his relationship with Josh as soon as tonight, but Ned may not be so forgiving. It didn't have to be this way. With a little knowledge about his brain, Paul could have reached his insight about getting someone to help him with offshoring some work without the collateral damage. Paul needs to know how to stay cool under pressure in a new way. He needs to learn to manage his expectations, in particular the expectations of positive rewards.

WHAT TO EXPECT FROM AN EXPECTATION

Up until now in this act we've focused on managing the threat response, because it is more common and stronger than the reward response. And who needs to learn how to handle the emotions that result from a good meal or great conversation? However, positive situations can sometimes send you off kilter, too. In a game of poker, if you are dealt a pair of aces, the best possible cards you can get, it's easy to get overexcited about winning the hand. This excitement about potentially winning creates a lot of arousal in your limbic system. While this high level of arousal might feel pleasant, the outcome is similar to negative arousal: there are fewer resources available for your stage, so you don't think as clearly. The result is you miss ways that you could still lose that normally you would easily notice. Mistakes made this way, both at the poker table and in life, can be expensive.

Paul's situation in this scene is like that of expecting to win with a pair of aces. It's not an *actual* positive reward that's thrown him here; it's the *expectation* of a reward. The expectation of a positive reward has quite an impact on the brain, changing not just your ability to process information, but even what and how you perceive. Expectations are also central in the creation of upward and downward spirals in the brain. They can take you to the peak of performance, or to the depths of despair. Maintaining the right expectations in life may be central for maintaining a general feeling of happiness and well-being. Creating just the right expectations is also an opportunity for your director to write the emotional script of your daily life rather than only reacting to challenges as they arise.

WHAT YOU EXPECT IS WHAT YOU EXPERIENCE

An expectation is an unusual construct, as it's not an actual reward, but rather a feeling of a *possible* reward. Whether you are seeing a delicious berry in real life or on your mental stage, or just expecting to see one, in each case the map for "berry" is activated, as is your reward circuitry.

Positive expectations are all about sensing that an event or item of "value" is heading your way. Value in the brain, of course, means something that will help you survive and reproduce. Primary rewards such as sugary foods and sex are generally tagged as valuable by the limbic system. You can also create your own maps for items or experiences that you decide are of value. You could choose to value good-quality shoes. In this case, like Carrie in *Sex and the City*, just walking past a shoe store might make you happy. In Paul's case, he has created a map of billions of interconnected neurons that represents the potential profit from this project. This map has become denser because he has thought about it, paid attention to it, and even talked about it, in this case to his wife in relation to a vacation.

Another example of a self-created map for something of value is when you set a "goal." When you set a goal, you make a decision that an end result is of value. As you think about this goal, or work toward it, you increase the expectations of a reward. Heading toward a goal can activate an overall *toward* state in the brain.

Your brain automatically orients toward events, people, and information that connects to what you have valued positively. In a paper called "The Neuroscience of Goal Pursuit," Elliot Berkman and Matthew Lieberman explain that "Participants in several social psychological studies have been shown to orient toward goal cues and engage in goal pursuit while being completely unaware of both actions." That's why when I decided to have children, I started to notice carriages, playgrounds, and kids' menus everywhere. This principle has been studied right down to the neuronal level. Scientists have trained monkeys to expect to see a specific object, say, a red triangle. Neurons in the monkey's brain for perceiving a red triangle light up *before* the triangle appears. The phrase "seek and you shall find" may have a basis in neuroscience.

Because expectations alter perception, this leads people to see what they expect to see, and not see what they are not expecting. When Paul's spreadsheet does not meet his expectations, he discards the data, assuming an error has been made. Josh thinks there's no food in the house and therefore doesn't notice opportunities for snacks. Michelle doesn't have that expectation, and sees a different world in the same refrigerator.

Unmet expectations often create a threat response, which I explain further later in this scene. Because the brain is built to avoid threat, people tend to work hard to reinterpret events to meet their expectations. It's all too common to see people make tenuous links between ideas that are not really linked, or discard important data that might disprove a theory. This sometimes comes with tragic results, from police officers accidentally shooting someone they expect to be armed, to one country invading another nation based on assumptions that later prove incorrect.

THIS WON'T HURT A BIT

Some scientists believe that expectations can explain the placebo effect. In one study by Dr. Don Price, three groups of volunteers with irritable bowel syndrome (who were fully informed of what was to come and, it is hoped, well paid) had a balloon blown up in their rectums. They were asked to rate the pain on a scale of one to ten. One group experienced the pain with no medication. These data are represented by the solid line of the graph below, with an average pain rating of 5.5 out of 10.

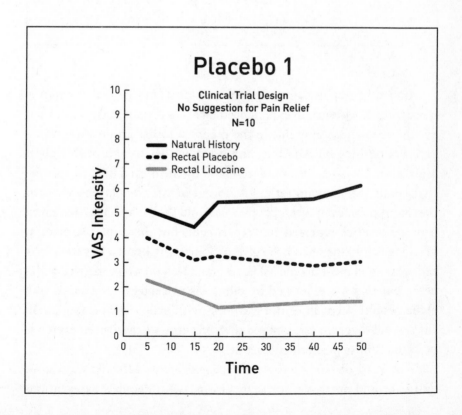

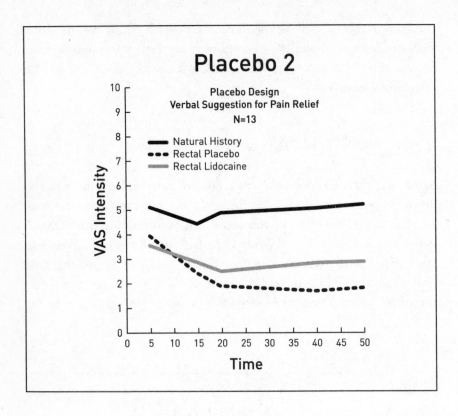

Another group was given lidocaine, a local anesthetic that removes most feeling. This group experienced an average pain rating of 2.5 out of 10, the bottom, gray line in the graph. Another group was given a placebo, nothing but Vaseline, and was told they were probably given a placebo. Their results are represented by the dotted line; they experienced an average pain rating of 3.5 out of 10. The placebo lessened perceived pain, even when people were told they may have been given a placebo. Price repeated the experiment, but this time the placebo group was told they had been given "Something known to powerfully reduce pain in most people." They weren't told they'd been given a placebo, but they were not lied to, either, as placebos do reduce pain in some people. What Price did was mess with people's expectations. In the second diagram you can see that this group experienced even less pain than those given the lidocaine.

This kind of study has been repeated many different ways now, and over and again we see that messing with people's expectations can have a remarkable effect on their perception. Professor Robert Coghill, a pain researcher at the University of Florida, designed an experiment in which people experienced intense pain on the leg via a calibrated heat pad. He then played with people's expectations to see what impact this might have on how they rated their pain. "Ten out of ten of our subjects had pain ratings go down when they simply just expected they were going to get a forty-eight-degree Celsius rather than fifty-degree Celsius stimulus," Coghill explains. In a paper called "The Subjective Experience of Pain: Where Expectations Become Reality," Coghill added: "Positive expectations produce a reduction in perceived pain that rivals the effects of a clearly analgesic dose of morphine." The right dose of expectations can be as powerful as the one of the strongest painkillers. Dr. Bruce Lipton's book *The Biology of Belief* explores this phenomenon in more detail.

Coghill wanted to know if the placebo effect involved fooling yourself. Was it just a "mind thing," or were there real changes in the brain? He looked at brain scans of people experiencing reduced pain through altered expectations. He found that when people expect a moderate level of pain, but instead get a strong level, it changes the brain regions normally responsive to pain. "We get dramatically reduced activation of a host of brain regions," Coghill explains. Expect something good, or bad, and it impacts the activation of brain regions the same as the actual experience would when generated in "reality."

THE NEUROCHEMISTRY OF UNDER-PROMISING

Expectations don't just affect the data you perceive, and change the activation of brain regions. They also have a strong impact on your neurochemistry. The best research for this comes from Professor Wolfram Schultz, at Cambridge University in England.

Schultz studies the links between dopamine and the reward circuitry. Dopamine cells sit deep within the brain, in the midbrain, connecting from there to neurons in the nucleus accumbens, and fire off in anticipation of primary rewards. Schultz found that when a cue from the environment indicates you're going to get a reward, dopamine is released in response. Unexpected rewards release more dopamine than expected ones. Thus, the surprise bonus at work, even a small one, can positively impact your brain chemistry more than an expected pay raise. However, if you're expecting a reward and you don't get it, dopamine levels fall steeply. This feeling is not a pleasant one; it feels a lot like pain. Expecting a pay raise and not getting one can create a funk that lasts for days. However, low levels of unmet expectations are something we all experience constantly: expect the stoplight to change and find it taking a long time, and your dopamine level falls, leaving you feeling frustrated. Expect the service at the bank to be fast but find a long queue: more frustration. Not only does dopamine go down in these instances, it also generates an *away* response, reducing prefrontal functioning. You might need to reappraise, perhaps saying to yourself, "This is good motivation to set up my Internet banking for this task." Do this and you will find frustration dissipating and a *toward* response taking its place.

Dopamine is the neurotransmitter of desire. Dopamine levels rise when you want something, even something as simple as to cross the road. (Dopamine is a driver of the reward response in most of the animal kingdom, too. At last we know the real reason the chicken wanted to cross the road; it was craving a burst of dopamine!) Put simply, dopamine is central to the *toward* state, to being open, curious, and interested. It is even tied to the act of movement itself. Parkinson's patients, having lost most of their dopamine neurons, have trouble initiating movement.

The number of connections made per second in the brain is also tied to dopamine levels. A hit of cocaine dramatically increases dopamine levels, with people chaotically jumping from idea to idea as the number of connections per second increases. When dopamine levels are too low, the number of connections per second in the brain falls. The movie *Awakenings*, with Robin Williams and Robert De Niro, illustrates the story of a patient who goes from comatose to manic after being given a dopamine-producing agent, L-dopa. When the L-dopa is stopped, the patient plummets back into a comatose state.

The dopamine cells in the nucleus accumbens connect to many parts of the brain, including the prefrontal cortex, where the right levels of dopamine are critical for focusing, as we learned in act 1. According to Amy Arnsten, you need a good level of dopamine to "hold" an idea in your prefrontal cortex. Positive expectations increase the level of dopamine in the brain, and this increased level makes you more able to focus. This makes sense intuitively: teachers know that kids learn best when they are interested in a subject. Interest, desire, and positive expectations are slight variations on a similar experience, the experience of having an increased level of dopamine in the brain.

Paul's dopamine level plummeted when his profit expectations were not met. He experienced a sudden reduction in desire to do anything significant. He wanted to do mundane tasks that his assistant should have been doing. He also experienced a drop in the number of ideas his brain processed per minute and reduced overall activation of his brain. His brain was in an overall *away* emotional state, which made it harder for him to think about complex problems such as outsourcing his coding. Josh experienced a similar challenge. He was anticipating the reward of seeing a friend, and when his father thwarted his plan, he became upset and angry.

In Paul's reduced dopamine state, he struggled to think through his situation. The project looked hopeless. He was starting on a downward spiral, where a low dopamine level can lead to an even lower dopamine level. It wasn't until he became interested in something—in his daughter's day at school—that his dopamine level rose to a point where he could start experiencing a *toward* response. Then, when he had his own insight about outsourcing, he became excited, and was off again, taking action. The unexpected insight increased his dopamine level. Each positive new connection he made—finding possible suppliers, for example—increased his anticipation of further rewards, increasing his ability to make new connections. He had shifted into an upward spiral.

Michelle also found herself in an upward spiral in this scene. She was already in a positive state of mind, and when her grade for her artwork was better than expected, her mood improved even more. This upward spiral allowed her to see a possible solution to her dad's challenge, where he could see only problems. She even saw more options for food in the fridge than Josh had.

This upward spiral sounds suspiciously like it might partly explain why people perform better when they are happy. Lots of research has been done, such as by Barbara Frederickson from the University of

North Carolina, showing that happy people perceive a wider range of data, solve more problems, and come up with more new ideas for actions to take in a situation. The link between expectations, dopamine, and perception may explain why happiness is a great state for mental performance. Perhaps the elusive search for happiness is actually a search for the right level of dopamine. From this perspective, to create a "happy" life perhaps you should live a life with a good amount of novelty, create opportunities for unexpected rewards, and believe that things are always going to get slightly better.

CREATING THE RIGHT EXPECTATIONS

Whether your goal is to be eternally happy or just to improve your performance at work, clearly it's going to be useful to manage expectations well, to create the right level of dopamine. To be clear, I am not an advocate of consuming L-dopa, cocaine, or any other substance that induces a greater dopamine level. The best way to manage your expectations (without any side effects) is to start to pay attention to them, which means activating your director. Managing your expectations is also an opportunity for your director to be more proactive, setting the scene for good performance in the future rather than just sorting out problems when things go wrong.

Unmet expectations are one of the important experiences to avoid, as these generate the stronger threat response. "With any brain function, the important thing is firstly to minimize threat," explains Evian Gordon. "Only once threat has been minimized can you focus on increasing possible rewards."

Consciously altering what you expect can have a surprising impact. Imagine you are trying to get an upgrade for a long international flight. If you keep your expectations low, you will either be okay if you don't get the reward, or thrilled if you do. Whereas if you allow yourself to get excited about the possible upgrade, you will either have a terrible flight if you don't get it, or be only quietly happy, though not thrilled, if you do get it. When you step back and look at all the possible outcomes this way, it makes sense to minimize one's expectations of positive rewards in most situations. Keeping an even keel about potential wins pays off.

As well as making sure you keep your expectations low, another way to boost your mood is to pay additional attention to positive expectations you know will be met for sure. A colleague recently said, "I like to use the fact that I have a holiday coming up, even if it's months away, to help me be positive. If I focus on this, though it's not logical, I have learned this helps keep the doldrums away." Choosing to focus on things always getting a little bit better, even with evidence at times to the contrary, helps you maintain a good level of dopamine.

Great athletes know how to manage their expectations. They don't get overexcited about possibly winning, as this ruins their concentration. And if they are worried about losing, they try not to expect that, either. Managing your expectations in any way requires, as with labeling and reappraisal, a strong director. When you can stop and notice your own mental state, you have the capacity to make choices about different ways of thinking. Great athletes observe the flow of their attention and make subtle changes to where their attention goes. Their director might notice expectations getting too aroused, and choose to dampen excitement, pushing the brain to stay focused on the moment instead. To use your director well you need to be able to find it. One of the best techniques for this is simply to pay more attention to your own experience, including watching how expectations alter your state of mind.

Let's look at what Paul might have done differently here if he had had a strong director and had been able to manage his expectations even in his challenging situation.

WHEN EXPECTATIONS GET OUT OF CONTROL, TAKE TWO

It's 3:00 p.m. Paul is back at his desk, thinking about how he's going to deliver the new project. He agreed to the tight deadline, but asked for two days to create a detailed project plan.

He goes to open a spreadsheet, then stops to think about his mental process. He doesn't feel that he is in the right state of mind for this kind of work. He doesn't know why just yet, but a quiet signal is telling him to think about how to approach this budgeting task. He decides to take a brief walk to the store to get some milk, knowing he will have time to reflect along the way. On his walk, he remembers how excited he has been in the past about good profits, and how this excitement got in the way of clear thinking. He pushes aside the urge to get excited about the project, deciding not to give the thought any attention. He was just about to call a few of his suppliers to share the good news, but decided this might not be a good idea, in case expectations are not met. Paul's director can pull back the wrong actors just as they are about to go onstage, which is the easiest time to stop them.

Paul returns home and opens a spreadsheet to cost out the project. He works out that he'll need all of his contractors, and that they'll have to work more hours than usual. They will charge him extra, as they will have to put more staff on his project. He plugs in the numbers and scrolls down to see the profit figure at the bottom. It's in the negative. Paul knows he could get upset by this, but he vetoes his attention going down that path, telling himself he knows this is a first attempt at the pricing. There are probably ideas he hasn't thought of. He gets up and goes to the kitchen to get a snack, thinking that some glucose might help him make new connections, and maybe his unconscious will have an insight if he quiets his brain.

Josh comes in to look for food, and Paul uses the opportunity to explain his challenge, hoping that another perspective might help get around an accidental impasse. As he hears himself explaining the challenge out loud to Josh, he has an idea. The insight pops into his mind as he talks about there being no other way of doing this job with a profit. His director, observing his sensation as he said these words, realized how silly the words sounded; there's always another way. Speaking aloud about complex ideas can be a way of seeing your own thinking more clearly. Paul sees that this could be a great chance to try outsourcing his coding to someone in a cheaper country. In his more positive state, Paul maintains an "open mind" about the idea. He doesn't resist it, even though the idea is uncertain. A small amount of uncertainty is easier to take when you are in a *toward* state. Paul searches the Web for software suppliers in India and finds many. He gets a quick and positive response from a possible supplier. When Josh asks about going to the store to buy a snack, Paul asks his son about his homework, which he is pleasantly surprised to learn Josh has already done. In his happy state Paul gives

Josh some money and watches as he bounds out the door, happy to be seeing a friend, as expected.

When Michelle comes home, they chat about her successful day at school. Paul gives her positive feedback that leaves her beaming. She offers to cook dinner for everyone, and Paul says he will get takeout instead, so they have more time to hang out together. It's been a great day.

A FINAL WORD ON ACT 2

As we wrap up the second act, you now have three specific techniques for staying cool under pressure. Each of them requires activating your director and getting focused on the present, which will increase the space on your stage. For average emotional hits you can try labeling your emotions, which increases a sense of certainty and reduces limbic arousal. For stronger emotional hits you can reappraise, by changing your interpretation of events. This can increase both certainty and autonomy, while having a stronger dampening effect. And to help reduce future bursts of arousal, you can manage your expectations by being aware of what they are, and choosing new expectations in their place. Each of these techniques is improved by your having a strong director, and each of them builds your director further as you execute them. With these three techniques in hand—or, rather, easily accessible as maps in your brain—you have a great chance of staying cool under pressure, even in the most difficult of circumstances.

Surprises About the Brain

- Expectations are the experience of the brain paying attention to a possible reward (or threat).
- Expectations alter the data your brain perceives.
- It's common to fit incoming data into expectations and to ignore data that don't fit.
- Expectations can change brain functioning; the right dose of expectations can be similar to a clinical dose of morphine.

- Expectations activate the dopamine circuitry, central for thinking and learning.
- Met expectations generate a slight increase in dopamine, and a slight reward response.
- Exceeded expectations generate a strong increase in dopamine, and a strong reward response.
- Unmet expectations generate a large drop in dopamine level, and a strong threat response.
- The dynamic between expectations altering experience and impacting dopamine levels, helps generate an upward or downward spiral in the brain.
- A general feeling of expecting good things generates a healthy level of dopamine, and may be the neurochemical marker of feeling happy.

Some Things to Try

- Practice noticing what your expectations are in any given situation.
- Practice setting expectations a little lower.
- To stay in a positive state of mind, find ways to keep coming out ahead of your expectations over and again, even in small ways.
- When a positive expectation is not being met, practice reappraising the situation by remembering it's your brain doing something odd with dopamine.

ACT III

• • • • • • • • •

Collaborate
with Others

Few people work in isolation anymore. The capacity to collaborate well with others has become central to good performance in just about any endeavor. Yet the social world is also the source of tremendous conflict, and many people never master its seemingly chaotic rules.

The problems that occur between people could be reduced if there were a wider understanding of some of the basic needs of the brain. Along with the need for food, water, shelter, and a sense of certainty, there are "social needs," which, if not met, create a sense of threat that can quickly devolve into conflicts between people.

In act 3, Emily discovers just how much the brain needs social connections, and recognizes the surprising importance of feeling safe among friends. Paul discovers that a sense of fairness drives a lot of behavior, and learns to keep this under control in himself and a colleague. Emily finds out that a sense of status is a far bigger need than she ever expected, and discovers sustainable ways of increasing her own status without other people feeling threatened.

SCENE 10

● ● ● ● ● ● ● ●

Turning Enemies
into Friends

t's 2:00 p.m. Emily has just finished getting the sustainability conference approved at lunch. Now back at her office, she picks up the phone
and presses a button she programmed with her usual conference line
number, and joins a conference call with her team just in time. Without
the need to find the number consciously, Emily conserves her attention
and uses a moment to focus on the present, to reflect, and to activate her
director. She notices it takes longer to bring an idea onto her stage than
it did several hours ago. She tries to find a word for her state of mind
and comes up with *frazzled*. Naming her state calms her down. She also
notices a nagging discomfort in the back of her mind, but can't quite
name it. All of this thinking happens in the few seconds as she waits for
the recorded message from the conference call system to roll through.

Colin and Leesa are on the line when she connects. They abruptly
end a conversation as Emily joins in, which leaves an uncomfortable
silence. The three of them used to be peers, often working on a conference together till late at night. Emily wonders what it will be like to
manage her old buddies, and she senses her anxiety rising. She tries to
find a way to reappraise the situation but doesn't have the focus, and
moments later she is distracted further by Joanne, whom she hired this
morning, arriving on the line.

Emily tries to get her thoughts together by focusing on being organized. She presents the agenda: to decide who will run the sustainability conference, to introduce Joanne, and to plan how the team will meet regularly. She hopes she will be able to generate a feeing of their all being a "team," the sense of community she felt when she managed a group of people working on a previous conference. Yet with everyone spread across the country, they will rarely meet in person, and everyone has their own conferences to focus on. Emily also briefly worries about introducing a new person to the group when there is a feeling of competition already. This quiet alert signal tries to make itself heard but fails to gain much attention.

"Everyone, I want to introduce you to Joanne, who is taking over the conferences I used to run," Emily says without pausing for air. "I chose her because she has run other large conferences successfully." She thinks she hears Leesa sigh, but she isn't sure.

"Nice to meet you all," Joanne responds, and then it's back to the agenda. Emily says she wants to choose someone to run the sustainability conference, and the line goes silent.

"Colin," Emily says, "you've worked with me the longest. Who do you think would be best for the job?" Colin and Leesa have never gotten along, but Emily is still surprised by what happens next. "I don't think Leesa is right for a new event," Colin says, "as she loves 'beautifully organized systems' rather than complexity." He tries to inject humor into his tone, but he is the only one laughing. Colin continues, unaware of the sharp rise in arousal in Leesa's limbic system: "And it wouldn't be right to put a new person onto this summit; it's a big event."

"Not to be rude," Joanne interrupts, "but, Colin, this is the same size event as the ones I used to do."

"Colin," Leesa jumps in, "you're not the best person when it comes to managing numbers yourself." Colin knows Leesa is harking back to a conference that lost money.

"Leesa, don't attack Colin on my behalf," Joanne responds. "I'm just saying that I have run big events, so don't count me out."

Emily tries to bring the focus back to the conference, but it's hopeless. Under their polished surface, Colin and Leesa are like two alley cats hissing at each other. Emily decides to close the meeting early, hoping she can sort this out with the team one-on-one.

Emily is disappointed. She can't understand why everyone has be-haved this way. She is especially annoyed at Colin, whom she thought she could trust. "He should know better than to bring a new person into the fold with such a negative experience," she thinks to herself. "Doesn't he appreciate how hard it is to find good people?" This in-tense emotional experience has burned a strong memory in Emily's hippocampus and amygdale. When she sees or thinks of Colin in the future, she will remember this conference call. She makes a mental note to treat Colin differently from the friend she thought he was. She then thinks about Joanne, wondering if she might decide not to take the job after all. This thought generates a lot of uncertainty, making Emily feel even worse. It's been a difficult and confusing half hour.

Success in most jobs today requires a strong ability to collaborate with others. For some people who build their mental maps around logi-cal systems such as computers or engineering, the chaos and uncer-tainty of dealing with people can be overwhelming. But it turns out there are rules to successful engagements in the social world. One of the big rules being uncovered is that the social world is deeply im-portant to our moment-to-moment existence. As Matthew Lieberman says, "Four out of five processes operating in the background when your brain is at rest involve thinking about other people and yourself."

Emily has been blindsided by the brain's social nature. She doesn't know how closely attuned the limbic system is to the social environ-ment, or how easy it is for people to misread social cues. In the ab-sence of positive social cues, it's easy for people to fall back into the more common mode of human interactions: distrusting others. In this brain state, with the limbic system overly activated, a joke becomes a slight, a slight becomes an attack, and an attack becomes a battle. And that can be the end of productive, goal-focused thinking for as long as humans can hold a grudge, which is a long time indeed.

Emily knows the rules that underpin how to run successful con-ferences. These involve managing budgets, suppliers, advertising, and systems. Like a classical musician learning jazz, Emily needs to learn new rules for successfully collaborating with others. In this scene, she needs to become better at turning enemies into friends.

THE BRAIN IS A SOCIAL ANIMAL

If you were a wolf, large parts of your brain would be devoted to getting resources directly from the wild. You would have complex maps for interacting with the physical landscape: maps for sniffing out a distant meal and others for finding your way home in the dark. As a human, especially when young, you get all of your resources not from the wild, but from other people. Because of this, large amounts of human cortical "real estate" are devoted to the social world. If you work in an office, you could probably close your eyes and describe ten people around you, how important they are in relation to one another and to you, how they feel today, whether they can be trusted, and how many favors any of them might owe you. Your memories of your social interconnections are vast.

Social neuroscientists think of the human brain as having a social network, responsible for all your interactions with your social world, similar to other networks you have for seeing, moving, or hearing. The social brain network allows you to understand and connect with others, and to understand and control yourself. It involves regions already discussed in this book including the medial prefrontal cortex, the right and left ventrolateral prefrontal cortex, and the anterior cingulate cortex, the insula, and the amygdale. This social network is something we're born with. Newborn babies orient toward a picture of a face, above any other picture, when just a few minutes old. At six months, well before they can speak, infants experience advanced socially oriented emotions such as jealousy. There is a lot of evidence that the experiences people think of as the best and worst of their lives are not individual achievements but social experiences, such as starting and ending important relationships.

All of this means that social issues matter to the brain. A lot. In fact, some scientists now believe that social needs are in the category of primary threats and primary rewards, as essential for survival as food and water. In the 1960s, Abraham Maslow developed a now famous "hierarchy of needs," which shows that humans have an order in which their needs have to be met, starting with physical survival and moving up the ladder all the way to self-actualization.

Social needs sit in the middle. But Maslow may have been wrong. Many studies are now showing that the brain interacts with social needs using the same networks as it uses for basic survival. Being hungry and being ostracized activate similar threat and pain responses, using the same networks.

In this first of three scenes about the social world, Emily has run into the need for feeling safety with others, a fundamental desire to feel related and connected to the people around her. I call this a feeling of "relatedness." A feeling of relatedness is a primary reward for the brain, and an absence of relatedness generates a primary threat. A sense of relatedness is what you get when you feel that you belong in a group, when you feel part of a cohesive team. Emily used to experience relatedness when she ran the conferences herself, but it's not there in her team now.

MIRRORS IN THE BRAIN

The way the brain creates a sensation of connection and relatedness with others involves a surprising discovery about the brain, made only in 1995. Emily's conference call went wrong because people made mistakes about the mental state of the others on the call. It began with Colin's attempt at humor being misunderstood. He made a joke that in a face-to-face encounter would have sparked laughter. Without his face and body language visible, everyone misread his intent. People on the call were not connecting the way the brain is best at connecting, which involves being able to copy other people's emotional states and intents directly. The way the brain does this is through mirror neurons.

Discovered by Italian neuroscientist Giacomo Rizzolatti at the University of Parma, mirror neurons have opened up a rich new understanding of how human beings connect with others. Rizzolatti discovered that mirror neurons throughout the brain light up when we see other people do what is called an "intentional action." If you see someone pick up a piece of fruit to eat, mirror neurons in your brain will light up. These same mirror neurons light up when you eat a piece of fruit yourself.

One of the unusual aspects of these neurons is that they light up only if we see someone perform an action that has a specific intent behind it. Random actions don't have the same effect. In this way, mirror neurons seems to be the brain's mechanisms for understanding other people's intent—their goals and objectives—and, as a result, feeling connected to them. Christian Keysers, a leading mirror neuron researcher based in Holland, says, "Our brain seems to make sense of other people through shared circuits. When you witness someone else taking an action, it activates the same circuits in your motor cortex. Someone picks up a glass; your brain does the same. It's through this capacity that you get this intuitive understanding of other people's goals."

Research into autism by Mirella Dapretto at UCLA has yielded further clues to the importance of mirror neurons. People with autism are considered "mind blind." They don't accurately decipher what other people are thinking, feeling, or intending, which results in social blunders. Many scientists now think mirror neurons are connected to autism, and new studies do indeed show damage to mirror neurons in this situation.

Keysers explains the way mirror neurons provide a direct experience of another person's intent. "What happens is that when we witness others' facial expressions, we activate the same in our own motor cortex, but we also transmit this information to the insula, involved in our emotions. When I see your facial expression, I get the movement of your face, which drives the same motor response on my face, so a smile gets a smile. The motor resonance is also sent on to your own emotional centers, so you share the emotion of the person in front of you."

Here's the source of Emily's challenge. On the conference call, without faces to see, the group couldn't read one another's emotions. The more social cues that are stripped out of communication, the greater the likelihood that intent will be misread. Most of us have experienced the difficulty of an email being misunderstood, of words being taken out of context. "The more we can see each other, the better we can match emotional states," Marco Iacoboni, a UCLA-based mirror neuron researcher explains. "Real interaction activates more than video, which activates more than telephone, as we are reacting to visual input of body language, and especially facial expressions."

If people don't have social cues to pay attention to, they can't connect with other people's emotional state. Studies show that the opposite is true, too. An abundance of social cues makes people connect more richly, perhaps in challenging ways at times. For example, when there is an abundance of social cues, emotional information can travel swiftly between people in a type of contagion. Studies show that the strongest emotion in a team can ripple out and drive everyone to resonate with the same emotion, without anyone consciously knowing this is happening. The strong emotion gets attention, and what people pay attention to will activate their mirror neurons. In a similar way, the boss's emotions can have a flow-on effect to others, since people pay so much attention to the boss. You see the boss smile, and your brain starts to mimic the smile; then you smile; then the boss smiles back. It's a virtuous and upward cycle, with each person raising the depth of the other's smile through a mirroring function. Mirror neurons explain why leaders need to be extra conscious of managing their stress levels, as their emotions really do impact others.

On the conference call, the strongest emotion was Leesa getting upset. This created a similar experience for the others on the call. While seeing faces will help the brain mirror others' brain states, mirror neurons can also work without seeing a face, through auditory cues, especially for the stronger *away* emotional states, which are often easier to arouse.

Christian Keysers says, "If you want to collaborate well with others you have to understand what kind of state others are in." Mirror neurons are the brain's way of knowing what other people are intending, and what they are feeling. They help you determine what your response should be to that person, whether to collaborate or cause trouble.

FRIEND OR FOE

While collaboration is increasingly important in our ever-more-connected world, working against this is the rise of the "silo mentality," where people collaborate within their own division, unit, or team inside a larger group, but don't share information more widely. It may help to understand that this is just human nature: People naturally tend to form safe tribes with close colleagues and work well within these, and avoid people they don't know well. That's because collaborating with people you don't know well is a threat for the brain. Perhaps, after millions of years living in small groups, the automatic response to strangers is "don't trust them." In a world of scant resources, where people lived only to an average age of twenty, this survival strategy worked. Now this reaction may be unnecessary, perhaps even a burden, especially inside organizations that depend on teams of people who should be deeply interconnected.

Here's one big reason collaboration is difficult: just as the brain automatically classifies any situation into a possible reward or threat, it does the same with people, determining, subconsciously, whether each person you meet is either a *friend* or *foe*. Is he someone you want to spend more time with (walk *toward* if you see him on the street) or stay *away* from (cross the road if we see him coming). And here's the rub: people you don't know, and particularly people you don't know who are also a little different from you, tend to be classified as foe until proven otherwise. This gets to the heart of Emily's challenge during the conference call. Not only did people misunderstand one another, but there was a strong sense of the others being a threat, being foes not friends.

Friend

You use one set of brain circuits for thinking about people who you believe are like you, who you feel are friends, and a different set for those whom you view as different from you, as foes. When your brain decides someone is a friend, you process your interactions using a similar part of the brain you use for thinking about your own experience. Literally, when you feel safely connected to someone, hearing them speak is similar to thinking your own thoughts. Deciding someone is a friend also generates a *toward* emotional response, which provides more space on your stage for new ideas. When you observe someone who you think is a friend, you more accurately assess their emotional states—you feel their feelings. Your brain lights up with the pain that a friend feels. This doesn't happen with a foe. And when you watch a friend do a task, you unconsciously cheer them on with micro-emotions like a smile when they do well or a look of concern when they struggle. This is reversed when someone is a foe.

When you interconnect your thoughts, emotions, and goals with other people, you release oxytocin, a pleasurable chemical. It's the same chemical experience that a small child gets when he makes physical contact with his mother, from the moment of birth onward. Oxytocin is released when two people dance together, play music together, or engage in a collaborative conversation. It's the neurochemistry of safe connectivity.

In a paper published in *Nature* in June 2005, a group of scientists found that giving people a spray containing oxytocin increased their level of trust. The paper reports that in nonhuman mammals, "oxytocin receptors are distributed in various brain regions associated with behavior, including pair-bonding, maternal care, sexual behavior and normal social attachments. Thus, oxytocin seems to permit animals to overcome their natural avoidance of proximity and thereby facilitates approach behavior." Our animal instincts seem to naturally cause us to withdraw and treat others as foes, unless a situation arises that generates oxytocin. This phenomenon makes sense: it explains why facilitators and trainers insist on "icebreakers" at the start of workshops, and why "establish rapport" is the first step in any counseling, customer service, or sales training manual. New research does complicate the story of oxytocin as the "trust drug" though. It appears to play a more general role in increasing all types of approach-related social behavior, including anger or jealousy. In short, while oxytocin can increase in-group trust, it can also promote aggression toward out-groups.

Research within the positive psychology field shows there is only one experience in life that increases happiness over a long time. It's not money, above a base survival amount. It's not health, nor is it marriage or having children. The one thing that makes people happy is the quality and quantity of their social connections. Daniel Kahneman, from Princeton University, did a study in which he asked women what they most liked to do. Surprisingly, connecting with friends was at the top of the list, above being with their partners or children. The brain thrives in an environment of quality social connections, of safe relatedness. Happiness is not just a good dose of dopamine, but a nice oxytocin buzz, too.

FRIENDS WITH BENEFITS

Having many positive social connections doesn't just increase your happiness; it can also help you perform on the job, and even live longer. The late John T. Cacioppo, who spent much of his career as a professor at the University of Chicago, studied the way human beings function socially and the impact the social world has on brain functioning. He led a study of 229 people between 50 and 68 years old, finding a 30-point difference in blood pressure between those who experienced loneliness and those with healthy social connections. Loneliness, the study showed, could significantly increase the risk of death from stroke and heart disease. As Cacioppo tried to understand the data, he realized that loneliness might be more important than society generally realizes. "Loneliness generates a threat response," Cacioppo explains, "the same as pain, thirst, hunger, or fear." Being connected to others in a positive way, feeling a sense of relatedness, is a basic need for human beings, similar to eating and drinking. For those of you who think that "hell is other people," remember that social isolation is not the brain's desired state. Having friends around you reduces a deeply ingrained biological threat response. Naomi Eisenberger, a social neuroscientist at UCLA, finds in her studies that increased social support also serves as a buffer against potential stresses by reducing reactivity to other threats. "I found that the greater levels of social support people said they had, the less sensitive they were to things like rejection." Eisenberger explains: "They seem to be less stress responsive. They even produced less cortisol." With less threat, people with good social support networks have more resources for their stage, for thinking, planning, and regulating their emotions, for example.

Surrounding yourself with friends not only helps you think better, it also enables you to see situations from novel perspectives, by "looking through other people's eyes." Friends provide a helping hand for that all-important but cognitively expensive emotional regulation tool, *reappraisal*. In the same way, having people you trust around can also help bring about insights, by broadening thinking and helping you to see your own thinking. This is all much more likely when people see each other as friend, not foe.

Having friends helps you change your brain, because you get to speak out loud more often. One experiment showed that when people repeated out loud what they were learning, the speed of their learning and their ability to apply that learning to other situations increased. When you speak to someone about an idea, many more parts of your brain are activated than just thinking about the idea, including memory regions, language regions, and motor centers. This is a process called spreading activation. Spreading activation makes it easier to recall ideas later on, as you have left a wider trail of connections to follow.

Need more evidence for the deep value of relatedness? A 2012 study showed that giving people even minimal social links to another person or group increased their motivation and improved their performance. A meta-analysis in 2010 of over 148 studies showed that people with stronger social connections had a 50 percent greater chance of survival than those with weaker social connections. And a 2011 study showed that people who feel they have higher levels of social support at work are literally less likely to die during any time period. Clearly, it pays to have quality social connections in your life.

Right now the introverts might be squirming, thinking I have lost my mind and that the worst thing they could imagine is spending more time with random humans. It is true that some people are energized by more social connections while others find them challenging. That doesn't mean that introverts should have few social connections; they just need different types. For introverts, they need fewer novel social connections and fewer surface connections. However, an introvert with only one close friend they feel safe with is likely to benefit from having more close friends they feel safe with. Although they need a different type of social connection, introverts will still benefit from the increased relatedness.

Foe

Recently I was invited to a party by a friend in New York City. I didn't expect to know anyone there, so I arrived late to be sure my friend would be there. When I arrived, he wasn't there. In theory I should have felt fantastic—the people at this party looked like the kind of people I would like, it was a beautiful loft, there was good music, nice food, and lots to drink. But I didn't know a soul, and because of that, my threat level was off the charts. To my brain, I had walked into a room full of foes. Five long minutes of trying to look calm later, my friend arrived, and my threat level went down dramatically. He introduced me to a few people, and I noticed with each new introduction how my threat level decreased. After an hour I had about six groups of people I could talk to, and it ended up being a lovely evening. This situation was a powerful reminder of how big an impact the foe response can have when you are surrounded by even falsely perceived foes.

When you sense someone is a foe, all sorts of brain functions change. You don't interact with a perceived foe using the same brain regions you would use to process your own experience. One study showed that when you perceive someone as a competitor, you don't feel empathy with him or her. Less empathy equals less oxytocin, which means a less pleasant sensation of collaboration overall.

Thinking someone is a foe can even also make you less smart. Kevin Ochsner explains: "Imagine trying to do business with someone who you've had conflict with in the past. Perhaps you keep getting distracted by thoughts of them being attracted to your girlfriend. Thinking of them as an opponent changes how you interact. You're focused on how you interact with them, instead of dealing with the business at hand." In this case your brain is trying to solve two different problems: how to deal with a foe, and how to do some business. Yet, as we know from act 1, multitasking is difficult. Neither goal is given enough resources, and mistakes are made. And mistakes generate yet more threat responses in the brain.

When you think someone is a foe, you don't just miss out on feeling his emotions; you also inhibit yourself from considering his ideas, even if they are right. Think of a time you were angry with someone. Was it easy to see things from his perspective? When you decide someone is a foe, you tend to discard his ideas, sometimes to your detriment.

Deciding someone is a foe means you make accidental connections, misread intent, get easily upset, and discard their good ideas.

In the new team structure, with Emily at the helm, Leesa decided during that first call that Colin was a foe; Colin decided that Leesa was a foe; both Colin and Leesa decided that Joanne was a potential foe; and all of them thought Emily was a foe. And Joanne probably just wanted to get the heck out of there. It's likely this occurred because they were all emotionally charged at the prospect of meeting a new person. Emily's big mistake was not recognizing how critical the social environment is. She didn't know that she needed to defuse the natural "foe" state before getting her new team to do some difficult thinking.

All of this explains the importance of the concept of inclusion, an idea that has taken hold in organizations over the last decade or so. There are many studies pointing to the fact that teams in which people feel included and feel they can speak up perform better. There's solid science to this: put simply, when people feel safely connected to others, when there is good relatedness, they think better. We need people to at least feel like friends, not foes, so that everyone can do their best work.

GOING FROM FOE TO FRIEND

While this foe response might seem like a scary monster best avoided altogether, it turns out to be easy, in most situations, to turn things around (provided you are not dealing with a deep foe response, such as a centuries-old ancestral grudge).

A handshake, swapping names, and discussing something in common, be it the weather or traffic, can increase feelings of closeness by causing oxytocin to be released. Emily launched straight into her meeting without the chance for people to connect on a human level. Allowing a few minutes for them to "get related" may have made all the difference. Think of these activities as shared experiences. When we have a shared experience with someone, we tag them more likely as a friend. This was shown in one particular study, which illustrated that increasing intergroup contact reduced prejudice toward outgroup members.

This can include any kind of experience, assuming we don't have a negative interaction with that person. This approach generally works well with turning strangers into friends. What about when we have someone who already thinks of us as a foe, perhaps someone from another team who you fight with for budget? In this case, when the brain notices that we have competing goals with someone, we tag them as a foe. To offset this, we need to find the shared goals. When we find a shared goal, especially something tactical like a task to complete this week, we turn enemies into friends. "The enemy of my enemy is my friend" is an old saying that is explained by this idea. A number of studies illustrate this idea, particularly coming out of Jay Van Bavel's lab at New York University, where he showed that switching up people's affiliations with one another—meaning having shared goals versus competing goals—quickly altered their interpretations of events and other people. While shared experiences make a difference, shared goals are really the driving force of relatedness. When you're trying to collaborate with anyone, start with a shared goal, and everything after will be easier.

Going from an automatic foe to a friend is not so hard. You probably do it many times a week without noticing. And creating shared goals with people, when you remember to do it, is highly impactful and not that difficult. Unfortunately, going from former friend to foe can be all too easy, even after years of positive interactions. This happened between Emily and her old work buddies, now that Emily was the boss and seen as the "enemy." Emily also decided not to trust Colin anymore, making a mental note about this after the meeting, yet they had worked well together for years. Because the away emotions, such as being upset with someone, are strong, going from friend to foe can be an intense experience.

It's hard enough for Emily's team that they won't meet in person often. What about collaboration between people from different cultures, people unlikely to meet at all? In this case the automatic foe response might need to be mitigated by dedicating social time in other forms. Perhaps ensuring the people forming teams share personal aspects of themselves via stories, photos, or social-networking sites. Some organizations set up defined buddy systems, or mentoring or coaching programs, all of which foster a sense of relatedness. A study by the Gallup organization showed that companies that encourage water-cooler conversations exhibit greater productivity. Increasing the quality and quantity of social connections (up to a point, of course) is likely to improve productivity, as more people find themselves surrounded by fewer foes, first and foremost, and then more friends as well. We simply perform better when we have positive relatedness with others. A great way to increase relatedness at work is to use video when having conference calls. As well as creating more relatedness, it helps with accurately reading social cues and can actually speed up the meeting through the use of hand signals and other nonverbal signs. For example, asking the group to "put your hand up if you can hear me clearly" over a live video feed is far more efficient than asking a large group of people the same thing verbally without video.

With all this in mind, it's time for a take two of Emily's conference call. Let's see how differently this could have gone if she had understood the importance of the social world.

TURNING ENEMIES INTO FRIENDS, TAKE TWO

It's 2:00 p.m. Emily has just finished getting the sustainability conference approved at lunch. Now back in her office, she picks up the phone and presses the button for her conference line number. She joins a conference call with her team just in time. She takes a moment to focus and observe her thoughts and internal state, to activate her director.

She notices it's taking longer now to bring an idea from the audience onto her stage than it did a few hours ago. She tries to find a word for her state of mind, and comes up with *frazzled*. Naming her state calms her down a little, but she also notices there's still something bugging her that she can't quite name.

Emily knows how delicate social situations can be, especially when people meet others for the first time. She pauses to pay attention to the issue bugging her, the impasse. She recognizes a pattern, buried deep within her limbic system. It's a weak connection that could emerge more clearly if she focuses on it. She puts her phone on mute to buy a moment to focus. In a couple of seconds an insight materializes. She realizes how important this meeting is, a first one for Joanne, and her first as boss. She senses that she has not done sufficient preparation to ensure that the call go well, and her agenda is wrong. She thinks about the main priority and recognizes that she needs to try to create a feeling of a "team" with the group before tackling anything hard together. She decides to run a more informal call and not be too ambitious about the conference. She thinks through this scenario, activating billions of circuits in the process, in the few seconds her line is on mute, then rejoins the call. She feels more certain now, and has resolved the impasse. Emily's brain is now in an alert but calm state, perfect for noticing subtle signals.

Colin and Leesa are on the line already, and they end a conversation as Emily joins the call. She senses an uncomfortable silence. If she hadn't paused and had her insight, she may have reacted badly. "Are you two plotting against me again?" she quips in a funny tone, making everyone laugh. She knows it's important to create connections between them all, as they used to have working together.

Joanne joins the call moments later. Emily explains that there is no formal agenda for the conference call, except for them to get to know one another and talk about how best to work as a virtual team. Emily asks the group for ideas about how they could get to know each other. She wants the group to think about their thinking a little, to activate their own directors. Leesa pipes in, saying it would be good for everyone to introduce themselves and share some of the more successful conferences they've worked on. When Emily first arrived on the line, Leesa and Colin had been talking about how anxious they felt in this new and uncertain team, with an unknown team member joining, and what Emily would be like as a boss. All this contributed to a sense of threat for Leesa and Colin. But by being given the opportunity to come forward with an idea, and then making a choice, Leesa shifted herself into a *toward* state. Joanne adds an idea to send around photos of themselves with their families. Leesa discovers that Joanne has children the same age as her own, and that the two women obtained the same degree. Leesa reclassifies Joanne as like herself. Conversations with her from now on will be more like speaking to herself—an open channel.

Emily goes last. She says she is new to managing people at this level and asks the group what they would like from her. Ideas pour out, with one idea prompting another, and a few themes emerge: People want open communication, trust, and respect. They also want to have fun. The group is resonating, generating good levels of oxytocin. The experience will be tagged as a pleasant one, and they will all look forward to the next call.

Colin asks if Emily has sign-off on the new conference. She almost announces that she wants to choose a leader for the sustainability conference, but in her less threatened and quieter mental state, she notices her own fear that the conversation could go wrong at this point. She says she will speak to them individually to get their thoughts, but Colin jumps in and says he thinks Leesa should tackle a big conference, as he was given the last one. Leesa asks Joanne if maybe they could do the event together, to get Joanne up to speed faster. The girls agree to collaborate, both knowing how much more fun it can be, and how much smarter they feel when they work as a team. The decision is made on the spot, and the group schedules the next meeting to start planning.

This second-take scenario and the opening one diverge for the briefest of instants. The positive changes occur with Emily noticing her mental experience, and with her having explicit language for the social world. As she develops this language further, she has an even greater chance of maximizing her performance.

Surprises About the Brain

- Social connections are a primary need, as important as food and water at times.
- We know one another directly through experiencing other people's states ourselves.
- Safe connections with others are vital for health, and for healthy collaboration.
- People are classed as friend or foe quickly, with foe as the default in the absence of positive cues.
- You need to work hard at creating connections to create good collaboration.

Some Things to Try

- Anytime you meet someone new, make an effort to connect on a human level as early as possible to reduce the threat response.
- Create an in-group with the people you work with by sharing personal experiences, or create an opportunity for any kind of shared experience.
- Create shared goals with people you feel you might be in conflict with. Ideally, these goals should be tactical short- to medium-term focuses.
- Actively encourage people around you to connect on a human level to create better collaboration.

SCENE 11

• • • • • • • •

When Everything
Seems Unfair

The phone rings. Paul lets it ring longer than he normally would. The day has been rough, and his limbic system is on high alert. He picks up the phone, hoping for a wrong number. No such luck. It's Ned.

Paul and Ned worked for the same consulting firm for several years before each deciding to go solo. They toyed with the idea of partnering up, but decided to run independent firms and collaborate instead, with Paul designing software strategies and Ned providing the more detailed software coding. This system worked well until today. What began as an ill-considered email from Paul about Ned not being involved in this new job has built into a torrent of emotion for them both, fracturing an otherwise deep bond. Given their long history, Paul wants to turn Ned from a foe back into a friend, but is not sure how.

"We need to talk," Ned says.

"I'm sorry about the emails," Paul jumps in, hoping an apology might be all that's needed. "You deserve more than that after all these years."

"That's what I want to talk to you about," Ned replies.

"Sure. But the thing is, I have been over and over the costs of this project, and though it's a big one, the margins are too tight. I am up against offshore vendors, and have to send the coding overseas to be in the black at all."

"I understand all that." Ned pauses for a moment. "Look, both of our emails were foolish, but that's not why I am calling. I just don't think what you're doing is fair. I have saved your butt many times over the years, working around the clock several times. You probably wouldn't still be here without my help. Why can't you cut me in somehow? It's a big project and I'm sure I could be of help somehow."

Paul is lost for words. He knows he was wrong to promise Ned any work before confirming the pricing. And he knows Ned must be disappointed. But he can't just cut Ned in. He would lose money, and that's something he's not willing to consider. Paul starts to feel that it's Ned who's being unfair here. The arousal in Paul's limbic system rises, including in his insula, which becomes active with strong emotions such as disgust. Doesn't Ned know how hard it was to get this project? Paul finds himself getting more upset by the minute, his emotions having automatically mirrored Ned's over the course of the call. He manages to eke out an apology through gritted teeth, trying hard to suppress his emotions. "I'm sorry, Ned. There's nothing I can do, really. But if I see a way to involve you in the project, I promise to keep you in mind."

Paul gets off the phone, sensing his relationship with Ned will never be the same. He can't pinpoint it, but Ned's asking to be cut in disturbs him deeply. He thinks it's so unfair for Ned even to bring the conversation up.

Paul hears Michelle turn on the television in the living room and gets out of his chair without thinking about it.

"Have you done your homework yet?" he yells across the room. Normally he wouldn't ask this kind of question, at least not in this way, but the arousal from the call with Ned has sent his director packing, making it harder for him to inhibit the wrong impulses.

"Dad, we agreed I have to do only an hour a day, and I could choose when I do it, anytime till eight thirty."

"Well, you know as it gets later you take longer. Why don't you get started now?"

"We made a deal, Dad. You can't change your mind now. And

Josh is just mucking around playing video games right now, too, you know."

"Oh, not you, too," Paul says, shaking his head.

"What? Why are you being so nasty today?"

"I'm not being nasty," Paul snaps back, "I'm your father and I have a right to ask about your homework."

"Well, just leave me alone, will you? It's not fair I have to put up with your bad moods from work."

"Okay, okay. Just make sure you get your homework done."

Paul's limbic system has been aroused by something big, something that happened as he tried to collaborate with others, something that happens surprisingly often anytime people work (or play) together. Paul doesn't know that *fairness* is a primary need for the brain. A sense of fairness in and of itself can create a strong reward response, and a sense of unfairness can generate a threat response that lasts for days. Just as Emily needed to change her brain to be more effective at turning foes into friends, Paul needs to change his brain to remember to maintain a sense of fairness with the people with whom he collaborates. As he learns to better manage fairness, Paul will find he can get far more done with less effort, and achieve his goals more easily as a result.

FAIR'S FAIR ————

You may find that once you prime your prefrontal cortex to look out for fairness issues they start to appear everywhere. Politics, to begin with, often involves emotional, even violent clashes, built on issues of fairness. While writing this book, I turned on the television to see a villager in Kenya shouting that she was willing to die to right the injustice of an unfairly rigged election. Fairness-generated emotions can run high in more mundane situations, too: the feeling of being "taken advantage of" by a taxi driver taking a longer route can wreck an otherwise great day, despite the relatively insignificant economic cost. It's the *principle* that counts. Think of people who spend enormous sums of money to "right wrongs" in court, with no obvious economic win other than "justice" or "revenge." We crave fairness, and some people spend their life savings and even their lives to get it.

FAIRNESS CAN BE MORE REWARDING THAN MONEY

Golnaz Tabibnia, an assistant professor at Carnegie Mellon University, studies fairness and the way people make judgments about it. "The tendency to prefer equity and resist unfair outcomes is deeply rooted in people," Tabibnia explains. One of Tabibnia's studies, in collaboration with Matt Lieberman, uses an exercise called the Ultimatum Game. In the Ultimatum Game, two people receive a pot of money to split between them. One person makes a proposal about how to split the money, and the other person has to decide whether to accept the proposal or not. If they don't accept the proposal, neither of them gets a reward. "'Inequity aversion' is so strong," Tabibnia explains, "that people are willing to sacrifice personal gain in order to prevent another person from receiving an inequitably better outcome."

Surprisingly, when people receive five dollars out of ten dollars, their reward center lights up more than when they receive, say, five dollars out of twenty. "In other words, the reward circuitry is activated more when an offer is fair than when it's unfair, even when there is no additional money to be gained," Tabibnia explains. Fairness, it seems, can be more important than money.

Tabibnia explains how this works in the brain. "There is a brain region called the striatum that responds when we get what's called a primary reward. The striatum receives rich dopaminergic input from the midbrain and is involved in positive reinforcement and reward-based learning. When people experience fair treatment, this circuit is activated. But when they experience unfairness, their anterior insula is activated. The reason this is interesting is the insula has been associated with disgust in previous studies, when you get a disgusting taste. Gustatory and social disgust are being processed in the same part of the brain—just as social reward and gustatory reward are processed in the ventral striatum. So it seems that these social reinforcers might map onto the brain in a similar way (at least in part) as more primary reinforcers."

Fairness doesn't intuitively feel like it is of the same importance as, say, food or sex. Because of this, many people don't tend to value fairness highly enough and, as a result, like Paul in this scene, are blindsided by the intensity of a fairness response. This is another example of Maslow perhaps being wrong. Society values survival needs such as food, well before social issues such as fairness. As a result, someone planning a day-long team meeting might pay attention to ensuring everyone has a good lunch break, but forget all about people's perception of fairness around how the day is organized. More and more research points to the insight that a sense of unfairness could be harder to handle than an empty stomach.

FAIR PLAY

Neuroscientist Steven Pinker has a theory about where this intense response to fairness comes from, which he's outlined in his book *How the Mind Works*. Pinker thinks that the fairness response has emerged as a by-product of the need to trade efficiently. In your evolutionary past, when you couldn't store food in the refrigerator, the best place to store resources would have been by giving "favors" to others. Resources were stored in other people's brains, as potential reciprocal snacks down the road. This mental exchange was especially important in hunter-gatherer days, when protein sources arrived intermittently: a bison felled by one person would be too much meat just for his family. To be good at this kind of trading you need the ability to detect "cheaters," people who promise but don't deliver. In this way, people with strong fairness detectors would have an evolutionary advantage.

These days, with fridges and bank accounts, you don't need to trust other people in such a primal way. Your fairness-detecting circuits are still there, but now they tend to get more of a workout in the form of leisure activities, such as the game of "cheat" played by kids, or Texas Hold-'em poker, now played by millions of adults the world over. These games provide an opportunity to flex your cheating and cheater-detecting muscles. While fairness in real life can generate a threat or a reward, detecting unfairness can be fun for the whole family.

WHEN IT'S JUST NOT FAIR

Let's dive deeper into the threat and reward response relating to fairness, beginning with the more common and stronger feeling of unfairness. Perceiving unfairness generates intense arousal of the limbic system, with all the attendant challenges this brings. As one example, because of the generalizing effect, accidental connections become easier: if you think one person is being unfair, everyone else may seem to be acting unfairly, too. In this scene, Ned's fairness issue came from thinking Paul wasn't taking into account their long history of helping each other. In an overly aroused state, Paul responded with his own sense of unfairness, mistakenly thinking that Ned was asking him to lose money on the project.

Many arguments between people, especially those close to us, involve incorrect perceptions of unfairness, triggering events that activate an even deeper sense of unfairness in all parties. This often starts by someone misreading one person's intent, being slightly mind-blind for a moment. The result can be an intense downward spiral, driven by accidental connections and one's expectations then altering perception.

Labeling may not be strong enough to bring a fairness response under control. You may need a more powerful tool, such as reappraisal. An important type of reappraisal here is to see the situation from the other person's point of view. But reappraisal takes a lot of resources, which makes it hard to do when unfairness kicks in. And seeing someone else's point of view when you have tagged him as a foe is hard, too. To manage an unfairness response, you might have to do it quickly, before arousal kicks in.

Since unfairness packs a hefty punch, it's easy to get upset by small injustices when you're tired, or when your limbic system already has a strong base load of arousal. You have to be extra careful in these situations. If you are kept awake by young children, it's easy to get cranky with a partner asking you for help. If you've had a bad day at the office, it's easier to get unnecessarily annoyed with a supplier who you think might be ripping you off, even though it might only be for pennies.

Fairness comes up a lot when dealing with children. "Do as I say, not as I do" is a statement parents wish they could use, but kids are finely attuned to fairness, even from an early age. Michelle felt insulted by being treated unfairly, and identified that Paul was treating her differently from her brother. With the teenage brain, small emotional hits can bring strong reactions. Prefrontal cortex functioning tends to shrink briefly as teens hit puberty, which explains why a ten-year-old may have better emotional control than a fifteen-year-old. Prefrontal functioning recovers in late teens and reaches an adult state only in the early twenties. (One theory for why the teen brain seems to go backwards for a while is that in the past, teenagers who did irrational things, such as having children, passed on their genes more than people who exhibited self-control.) Because of their poor emotion-regulation capacity, teens tend to feel the threats and rewards arising from fairness (and certainty, autonomy, and relatedness) very intensely. Perhaps this explains both the door-slamming arguments teens have with their parents, along with their influx into social justice projects.

JUSTICE IS ITS OWN REWARD

On the plus side, fairness is hedonically rewarding, activating dopamine cells deep in the brain the way a good meal or an unexpected bonus at work might do. When you experience a fair response, it's likely that serotonin, a neurotransmitter that puts you at ease, increases, although no studies have yet shown this directly. Prozac and Zoloft are antidepressant drugs that work by increasing the level of serotonin in the brain.

The feeling you get from a sense of fairness is one of connecting safely with others, so it's linked to relatedness. When you feel someone is being fair, there is also a feeling of increased trust. Studies show that a self-rated sense of trust and cooperation increase when people experience fair offers. Oxytocin levels increase in fair exchanges, too.

So an increasing sense of fairness increases your levels of dopamine, serotonin, and oxytocin. This generates a toward emotional state that makes you open to new ideas and more willing to connect with other people. This is a great state for collaboration with others. Yet so many structures inside organizations, especially large organizations, work against employees feeling a sense of fairness. The all-too-common complaints about pay, performance, and transparency are certainly linked to fairness. In the big downsizings of 2009, one firm's executives agreed to a pay cut of 15 percent, making a big deal that this was three times more than the 5 percent cut all staff were being asked to undergo, to help reduce layoffs. While a 15 percent cut meant thousands of dollars a year less pay for an executive, this didn't affect their bonuses, many in the tens of millions of dollars. You can imagine how employees felt about that. In a separate incident, there was furor at the time around bonuses at AIG, paid with government bailout funds to executives after the company lost billions, that nearly undid the global economy.

One interesting implication of fairness research is the idea that workplaces that truly allow employees to experience an increasing perception of fairness might be intrinsically rewarding. A 2012 meta-analysis of studies showed that employees' perceptions of unfairness at work can negatively affect both their physical and mental health. This may explain why people perform better in certain workplace cultures. I asked one executive I shared a car ride with why he had stayed at the same company for twenty-two years. "I don't know," he replied. "I guess it's because they always seem to do their best to do the right thing by everyone." Organizations trying to increase a sense of engagement could do well to recognize that people experiencing a sense of unfairness may get just as upset as being told they won't get to eat for a day.

Some organizational research published in *The Harvard Business Review* around corporate restructuring found that when people understood that the decisions were made fairly, the impact of the downsizing was dramatically lower. On the other hand, people who feel themselves to be treated unfairly by an organization can generate no end of complaints. Living in a world that appears unfair affects people's cortisol levels, their well-being, and even their longevity. No wonder so many people won't stay in corporate jobs when they think that their company isn't doing the "fair thing" for its workers, customers, or for the community at large.

There is one place you can go to experience a regular increase in the sense of fairness, and that's to work for social justice organizations that distribute food to the poor or generally serve underprivileged communities. When you right perceived wrongs, such as people being hungry when there's food being wasted two blocks away, you increase your sense of fairness. Organizations that allow people to take time on community projects are letting their employees experience informal rewards through a perception of increasing fairness. Many employees find this a deeply satisfying part of their job.

As an added bonus here, a study has shown that giving to others activates a greater reward response than receiving gifts of similar value. So sharing your time or resources or donating money might help you not only sense greater fairness but also feel better than you would receiving a gift yourself.

EXPECTING FAIRNESS

I propose that there may be a dynamic between fairness and expectations that explains some of life's more intense emotional experiences. This may be an interesting area for future research. If you expect someone to be fair with you and they are, you get a nice positive dopamine high, for two reasons: first, from your expectations being met, and second, from the fairness itself. Unexpected fairness should be even more pleasant, which explains why the "kindness of strangers" can feel so meaningful.

If, however, you expect someone to be fair with you and they are not, you get a double negative: a significant dopamine low from expectations not being met, and from the unfairness. This may explain why the arousal is so strong when someone trusted, say, a friend who you expect to do the right thing, is unfair to you. Now you've got the "perfect storm" of arousal, a triple whammy. This experience also has a word, and it's the visceral experience Ned was feeling: betrayal. A feeling of even mild betrayal can be a very intense experience.

ACCEPTING UNFAIRNESS

So fairness is a big driver of behavior, more than most people expect. Yet it's not as if people crumble in the streets when a taxi passes them by for a more attractive passenger. We have ways of managing unfairness, and how the brain does this is quite interesting.

When people deal with unfair situations, it's not just that they don't get a positive reward; it's more complex than that. Tabibnia studied situations where someone could choose to accept an unfair offer in the ultimatum game. For example, a poor graduate student might want to accept an offer of twenty dollars out of fifty. She finds that people are either overwhelmed by the insult and don't take, or feel snubbed but tempted to take the money anyway. "In this case," Tabibnia explains, "when people accept an unfair offer, it's not intrinsically rewarding. Instead, they down-regulate their emotional response. The insula is activated, but they override this response. At this point in the experiment, activity in the right and left ventrolateral prefrontal cortex (VLPFC) increases, and activity in the insula is reduced. The more likely the person was to accept unfair offers, the more the VLPFC was active, and the less the insula was active. It seems that the better you can regulate your emotions, the better you can accept an unfair offer." There's that all-important right temple again. Accepting unfairness requires tools such as labeling and reappraisal, which both require lots of resources for your stage to execute.

Neuroscientist Tania Singer, at the University of Zurich, has dug deeper into fairness, looking at the relationship between fairness and empathy. She had subjects play a game with two other players, who were in fact actors. One of the actors appeared to be a jerk, and the other, a collaborator. The two actors then get electric shocks (or at least they appear to; all the fun was taken out of psychology experiments after Stanley Milgram). In Singer's experiment, either the good guy or the bad guy gets the shock. The study showed that women share the pain of both the good guy and the jerk, whereas men share the pain only of the good guy, and their reward center gets activated when the bad guy gets zapped. "Punishment to unfair people is an important pressure that helps support fair economic transactions," Keysers explains. A sense of unfairness can be created if someone is not punished. Think of the outcry when an executive is let off with a fine for losing millions of dollars of investors' money, when someone else might be jailed for stealing a purse.

MANAGING YOUR FAIRNESS RESPONSE

The world is not fair, especially the business world, where dog-eat-dog can be a highly rewarded behavior. Being able to manage your response to unfairness puts you at an advantage over others. One way to do this is to label your emotional state when you sense arousal increasing. Whether it's unfairness, or uncertainty, or lack of autonomy or relatedness, being able to put words to why you are feeling a certain way will decrease limbic arousal and help you make better decisions. If labeling doesn't work, try reappraising, by looking at the situation from different perspectives.

On the flip side of this, if you perceive an injustice that you think should be righted, you could choose to allow yourself to feel a sense of unfairness. Choosing to be driven by these emotions may help you push past fears inherent in taking action to right a wrong. Just remember that strong limbic arousal is good for physical activity but reduces creative thinking. Letting yourself focus on the idea that an opponent in a football game has been unfair might help you run faster. But allowing yourself to focus on a sense of unfairness at work might drive you to make career blunders in the boardroom.

Clearly if Paul had understood the importance of a sense of fairness to the brain, he might have made different decisions this afternoon. Let's find out how this could have gone.

WHEN EVERYTHING SEEMS UNFAIR, TAKE TWO

The phone rings. "We need to talk," Ned says.

"I'm sorry about the emails," Paul jumps in. "You deserve more than that after all these years. I know you must think this situation is unfair, and I want you to have a chance to say what you think. Then maybe we can brainstorm a bit about how we can help each other more, if not on this project, then on others." Paul knows Ned is feeling unfairness.

"Okay . . ." Ned is disarmed by Paul's approach. He expected an argument. Paul listens while Ned tells him how upset he is, and how unfair it all is. He senses himself about to get upset at some of Ned's comments, but catches himself several times, choosing to reduce his rising emotions by naming them. One comment nearly trips Paul up, and he has to reappraise the situation, consciously remembering that Ned has always helped him out. Paul has to work hard for several minutes to hold back his own fairness response. It's a lot of effort, but it's worth it. Ned feels better having spoken about his issues, and without Paul getting "caught up" in a mirrored emotional response. The whole event reduced arousal on both sides instead of increasing it. Ned, in a better mood now, shares something he was thinking earlier but didn't want to share when he perceived Paul as a foe.

"Look, the thing is, Paul, you've not thought about the hard-coding side of this project. I have experience with this type of thing, and you've underquoted that part of the job to the client. Why don't I do a little consulting on the project, acting as your advisor? I don't need a lot, as I wouldn't be doing the coding, but I might save you a ton of money by getting the brief more accurate. Already I can see how I can make you a few thousand dollars here."

"That's not a bad idea," Paul replies. "And maybe having you there will allay the client's fears that it's just me, a small supplier. We might have more clout together."

They finish the call, agreeing to decide on a fair figure for the consulting work before the next client meeting so Ned can come along. They both leave the call relieved, but also pleased about their possible collaboration. Their open and trusting conversation has increased the levels of oxytocin in both of their brains.

Paul hears Michelle turning on the television in the living room. He thinks about getting up to check on her, and remembers that he made a deal with her about homework. He goes in to ask if she wants anything from the fridge. The effort is worth the look of surprise on her face. He brings her a drink and they watch a show for ten minutes, laughing at a kids' sitcom, enjoying a moment of bonding.

Surprises About the Brain

- A sense of fairness can be a primary reward.
- A sense of unfairness can be a primary threat.
- Linking fairness and expectations helps explain the delight of the kindness of strangers, as well as the intense emotions of betrayal from people close to you.
- When you accept an unfair situation, you do so by labeling or reappraising.
- Men don't experience empathy with someone who is in pain who has been unfair, whereas women do.
- Punishing unfair people can be rewarding, and not punishing unfairness can generate a sense of unfairness in itself.

Some Things to Try

- Be open and transparent about your dealings with people, remembering that unfairness is easy to trigger.
- Find ways to sense increasing fairness around you, perhaps by volunteering or donating money or resources regularly.
- Don't let unfairness go unpunished.
- Watch out for fairness being linked to other issues such as certainty, autonomy, or relatedness, where you can get intense emotional responses.

SCENE 12

• • • • • • • • •

The Battle for Status

t's 4:00 p.m. The conference call ended over an hour ago, the team in disarray. Emily tries to work on something else, but her mind swarms with unanswered questions from the meeting. She wants to make it all fit together, but all she sees are impasses. She deletes and files emails for a few minutes before her director kicks in, recognizing she is avoiding a call that needs to be made.

As she dials Colin, a quiet internal voice tells her to prepare mentally first, but anger drowns out this alert. She's still mad as hell at Colin for upsetting Leesa.

"I thought that would be you," Colin says.

Emily feels it's wrong to tackle the situation bluntly, but this feeling is overridden by a stronger emotion, that what Colin did was unfair. "Why did you do that?" she blurts out.

"Do what? I just made a joke and she lost the plot. Don't blame me. She's usually got a sense of humor. I've made worse jokes before and it's been fine."

"But you know this was different," Emily responds.

"Look, don't come down on me. It's not my fault," Colin replies. "Why are you attacking me when it's Leesa who went crazy? I did nothing wrong."

"Colin, I thought you were on my side here," Emily says. "I wanted you to run the big conference, but how can I give it to you when you've acted like this in front of the team? People will think I am playing favorites."

"I am on your side. What are you talking about?" he replies, getting exasperated. Colin's and Emily's ability to understand the position of the other has decreased as they have become more combative.

"So why did you act like such an ass?" Emily asks.

As Emily says these words, she can tell they are going to be taken badly, her subconscious predicting the result as she speaks, but it's too late to take them back. Yet she is still surprised by the intensity of Colin's response.

"Just because we used to work together doesn't mean you can treat me like dirt. You're not so perfect yourself," Colin replies, his voice low and slow to make his point, like a dog growling and baring his teeth.

"I'm sorry, I'm sorry, it's been a tough day for me, too, you know. I am not doing so great in my first few weeks as a boss here."

Emily isn't really sorry, and it's hard to fool a brain so well wired to read social nuances. Instead of stepping back from the edge as she hoped, Colin senses her weakness and goes on the attack. "Look, don't complain to me about your promotion. You're the one who wanted it. There's no way I am going to let the others take away my hard-earned position on this team. I have been here longest, so I deserve the big conference. You know there could be good money in bonuses, but it's not just that. It's what I deserve after what I have put up with here, and—"

Emily interrupts. "Sure, yes you've worked hard, but that doesn't make it automatic that you—"

"Don't lecture me," Colin interrupts, "I have been around a lot longer than you, you know."

Emily tries to back down, but the damage has been done. Their relationship, on solid ground after years of collaboration, is sinking fast just a week into her new role. She never imagined that managing people could be so difficult.

After another fifteen minutes talking in circles, Emily and Colin agree to let it rest and speak again in a few days. Emily gets off the phone and stares at her computer screen, feeling even more lost than before the call. She wonders what insight she is missing that might clarify Colin's issue for her.

Emily dials Leesa.

"I know you're doing your best," she starts off, consciously trying to be more "tactful" this time.

Leesa lets out a sigh, then starts. "You know, I really didn't want to attack Colin. But he attacked me in front of a new person. I couldn't just let that slip."

Emily tries to reason with Leesa, imploring her to call Colin to patch things up, but Leesa is adamant that it's Colin's job to do the mending.

Emily doesn't know what to do. It's true that they all used to laugh at Colin's jokes, but it's also true that Colin should have been more sensitive. Everyone seems to have been in the wrong here, but no one is willing to let the others know they were wrong.

"Leesa, what can I do to make this right?" Emily asks.

"Don't worry about it. Things will settle down, and we'll get on with business. We don't all have to be best friends to get the job done."

Leesa is only partially right. You don't have to be best friends with people to do good work. But working with a perceived enemy is uncomfortable, and errors can be easily made due to a lack of sharing information and other by-products of a high threat level between people. Emily has a big challenge on her hands. Most of her team has decided that everyone else is a foe. This foe response isn't just because she didn't set up the meeting well, though that is part of it. Something else happened on the call that generated a strong threat response. Colin, Leesa, and Joanne all perceived a threat to one of their most valued possessions: their status. When trying to clear this up, Emily threatened Colin's status even further.

Along with relatedness and fairness, status is another major driver of social behavior. People will go to great lengths to protect or increase their status. A sense of increasing status can be more rewarding than money, and a sense of decreasing status can feel like your life is in danger. Status is another primary reward or threat. Your brain manages status using roughly the same circuits used to manage other basic survival needs.

MAINTAINING THE STATUS QUO

The Doge's palace in Venice is one of the most lavish and ornate centers of power the world has ever seen. Much of it is still in good condition today. At the heart of the palace is an unusual room lined floor to ceiling with drawers for thousands of documents. The documents kept here for hundreds of years, while precious, didn't relate to money. At least, not directly. They listed the "status" of every person in the city. If you were a Venetian a few hundred years ago, one of these documents listed whose child you were, whose child they were, and how you were connected to royalty, merchants, or others of importance. The document implicitly defined where you would live, what you would eat, how much you would be educated, whether people trusted you, how much attention others paid to you, even how long you would live. Times haven't changed much. High status today, whether you are a pop star, top athlete, or CEO, comes with similar benefits that have a deep impact on quality of life. We just keep records differently. That's now the job of gossip magazines.

Status explains why people will queue for hours on a frosty morning to get a signed copy of a TV celebrity's new book (a book they might have no plans to read). Status explains why people feel good meeting someone worse off than themselves, the German concept of schadenfreude. One brain study showed that reward circuits were activated when people saw others as worse off than they. Status explains why people love to win arguments, even pointless ones. Status explains why people spend money on underwear from a designer fashion store, when similar clothes are available for a fraction of the price. It explains, at least in part, why millions of people today play online games with no obvious benefits except playing for points to raise their status compared to that of others. Status probably even explains how technology companies get thousands of people to work for free, doing work that computers can't do: they get people to compete with one another on tasks such as labeling a photograph.

Status is relative, and a sense of reward from an increase in status can come anytime you feel "better than" another person. Your brain maintains complex maps for the "pecking order" of the people surrounding you. Studies show that you create a representation of your own and someone else's status in the brain when you communicate, which influences how you interact with others. One study showed specifically that we are extremely accurate judges of where we fall on the social ladder.

Changes in pecking order brings about changes in how millions of neurons are connected. Colin has had to change massive numbers of circuits to relate to Emily in a new way as his boss, and some of these changes are still taking place in this scene. If you have ever been in a relationship in which one partner starts earning more money than the other for the first time, you have perceived these wide-scale changes in brain circuitry take place, which can bring some interesting challenges.

Organizations set up complex and well-defined hierarchies, and then try to motivate people with the promise of moving to a higher level within that hierarchy. One company I know won't let you face your desk toward the window until you move from a "band 40" to a "band 50" role, even though you might sit right next to someone who is a band 50. Marketing departments use two main levers to engage human emotions through advertising: fear, and the promise of increased status.

Despite attempts by corporations to make status about the size of your car or the cost of your watch, there's no universal scale for status. When you meet someone new and size up your relative importance, you might do so based on who is older, richer, stronger, smarter, or funnier. Or if you live in some Pacific Islands, based on who weighs more. Whatever framework you think is important, when your perceived sense of status goes up, or down, an intense emotional response results. People go to tremendous extremes to increase or protect their status. It operates at an individual and group level, and even at the level of countries. The desire to increase status drives people to achieve incredible feats of human endurance. And the drive for status is behind many of society's greatest achievements and some of its worst examples of needless destruction.

ON THE WAY DOWN

As with all the primary needs, it's the threat response that is stronger and more common with status. Just speaking to someone you perceive to be of a higher status, such as your boss, activates a threat response. A perceived threat to status feels as if it could come with terrible consequences. The response can be visceral, including a flood of cortisol to the blood and a rush of resources to the limbic system that inhibits clear thinking. Colin felt his status was threatened on the conference call because the group didn't acknowledge his seniority. The first words that Emily used in this scene only made this worse: the phrase "Why did you do that?" implies a person is in the wrong. Colin was already primed when thinking about the conference call to feel a status threat, so it was easy for Emily to accidentally make things worse. The intensity of Colin's reaction took Emily by surprise. She had no idea he felt a status threat in the first place.

Naomi Eisenberger, a leading social neuroscience researcher at UCLA, wanted to understand what goes on in the brain when people feel rejected by others. She designed an experiment that used an fMRI to scan the brains of participants as they played a computer game called Cyberball. The game harks back to the nastiness of the school playground. "People thought they were playing a ball-tossing game over the Internet with two other people," Eisenberger explained during an interview down the road from her lab. "They could see an avatar that represented them, and avatars for two other people. Then, about halfway through this game of toss between the three of them, they stop receiving the ball and the other players throw the ball only to each other." Whenever I tell this story to a room full of people there is an audible "ouch" from the audience. Being left out, being classed as "less than" others, is a universally painful experience.

This experiment generates intense emotions for most people. Eisenberger says, "What we found is that when people were excluded, you see activity in the dorsal portion of the anterior cingulate cortex, which is the neural region that's also involved in the distressing component of pain, or what sometimes people call the 'suffering component' of pain. Those people who felt the most rejected had the highest

levels of activity in this region." Exclusion and rejection is physiologi-
cally painful. A feeling of being less than other people activates the
same brain regions as physical pain. Eisenberger's study showed five
different physical pain brain regions lighting up under this social pain
experiment. Social pain can be as painful as physical pain, as the two
appear synonymous in the brain. Think of the drop in your stomach
when someone says to you, "Can I give you some feedback?" That
drop in your stomach is a similar feeling to walking alone at night and
sensing that someone is about to walk up and attack you from behind:
perhaps not as intense, but it's the same fear response. This discovery
about the brain explains why Colin reacted with the human equivalent of a dog baring its teeth and growling: his brain thought someone
was about to hit him.

Because of the intensity of the status-drop experience, many people
go to great lengths to avoid situations that could put their status at risk.
This aversion includes staying away from any activity they are not confident in, which, because of the brain's relationship to novelty, can mean
avoiding anything new. This can have quite an impact on the quality of
life. This is Gross's *situation selection* working against you.

With the threat response from status being so high, reappraisal in
these situations can be difficult, unless you catch (i.e., label and reappraise) the emotion early on, in the first few seconds, before it takes
over.

Colin's reaction to a status issue in this scene was to fight. He attacks Emily's status right back, spitting out the phrase "You're not so
perfect yourself." He also attacks Emily's credibility, pointing out that
she is younger than he. Being managed by a younger person may generate an automatic status threat if people don't actively take another
perspective (which means to reappraise), such as being interested in
learning about a younger generation.

Colin didn't just fight, he also had a flight response. While he didn't
run away physically, he did run away mentally: he ran away from
thinking. If he had stopped to think about the situation, he might
have recognized that you can't say on the phone what you might say
in person.

The threat response from a perceived drop in status can take on
a life of its own, lasting for years. People work hard to avoid being

"wrong" in a situation, from a simple mistake made on a document, to an error in judgment about a major strategy. Think of some of the big corporate mergers that have gone bad, and the executives who made the decisions avoiding any responsibility. People don't like to be wrong because being wrong drops your status, in a way that feels dangerous and unnerving.

When you decide you are right, the other person must be wrong, which means you don't listen to what she says, and she experiences you as a threat, too. A vicious cycle emerges. Leesa, adamant that Colin has to patch things up, feels she is "right." Colin no doubt feels the same way. Being "right" is often more important to people than, well, just about anything else, at the cost of not just money but relationships, health, and sometimes even life itself.

As well as sometimes taking on a life of their own, the other trouble with status threats is how easily they can occur, generating a strong threat even in minor situations. Say you are at a meeting with a colleague, and for the first time in your working relationship, he asks to follow up with you about a project. It's likely you will interpret his request as a threat to your status: Doesn't he trust you? Is he checking up on you? Your threat response could make you say something harmful to your career. Remember that the limbic system, once aroused, makes accidental connections and thinks pessimistically. Just speaking to your boss arouses a threat. If you manage someone, just asking how her day is going can carry more emotional weight than one might think.

Many of the arguments and conflicts at work, and in life, have status issues at their core. The more you can label status threats as they occur, in real time, the easier it will be to reappraise on the spot and respond more appropriately. The director has a big role to play when it comes to status. But be careful about trying to help another person see what is going on. Saying "that's just your status threat talking" isn't a great idea at a meeting.

ON THE WAY UP

I recently interviewed an international ballet dancer who used to be a member of the London Royal Ballet. She told me how she was often bored and frustrated as one of many dancers, even though she was in a world-class troupe. That all changed when she moved to a smaller, less known troupe in her home city, and now was the leading soloist. She explained, "Finally I am the highest paid dancer in the company. I am the one at the front of the room. The minute you're at the front of the room, there's no boredom at all. The focus is on you, the space is your space. You feel at the top."

Studies of primate communities show that higher-status monkeys have reduced day-to-day cortisol levels, are healthier, and live longer. This isn't just monkey business (sorry for the pun). There is an entire book, called *The Status Syndrome*, by Michael Marmot, which illustrates that status is a significant determinant of human longevity, even controlling for education and income. High status doesn't just feel good. It brings along bigger rewards, too.

Status is rewarding not just when you have achieved high status, but also anytime you feel as if your status has increased, even in a small way. One study showed that saying to kids "good job" in a monotonous recorded voice activated the reward circuitry in them as much as a financial windfall. Even little status increases, say, from beating someone at a card game, feel great. We're wired to feel rewarded by just about any incremental increase in status. Many of the world's great narratives (and some of our not so great television franchises) have status at their core, based on two recurring themes. These stories involve either ordinary people doing extraordinary things (giving you hope that you could have higher status one day) or extraordinary people doing ordinary things (giving you hope that even though you may be ordinary, you are basically the same as people with high status). Even an increase in *hope* that your status *might* go up one day seems to pack a reward.

An increase in status is one of the world's greatest feelings. Dopamine and serotonin levels go up, linked to feeling happier, and cortisol levels go down, a marker of lower stress. Testosterone levels go up, too. Testosterone helps people focus, makes them feel strong and confident, and even improves sex drive.

With more dopamine and other "happy" neurochemicals, an increase in status increases the number of new connections made per hour in the brain. This means that a feeling of high status helps you process more information, including more subtle ideas, with less effort, than a feeling of lower status. With the reduced threat response from the increase in positive emotions, you have plenty of resources for the prefrontal cortex to help you think on multiple levels. This means that with a perception of high status, there is more chance you can activate your director when you want to.

People with higher status are better able to follow through with their intentions more—they have more control, more support, and more attention from others. Being in a high-status state helps you make the connections your brain expects to make, which puts you in an upward spiral toward even more positive neurochemistry. This may well be the neurochemistry of "getting on a roll."

GETTING AND STAYING ON A HIGH

Maintaining high status is something that the brain seems to work on all the time subconsciously. You can elevate your status by finding a way to feel smarter, funnier, healthier, richer, more righteous, more organized, fitter, or stronger, or by beating other people at just about anything at all. The key is to find a "niche" where you feel you are "above" others.

If you video recorded a standard weekly team meeting in most organizations, you might be surprised to find that a large percentage of the words spoken are intended to edge an individual's status higher, or edge other people's status lower. This general bickering, the corporate equivalent of sibling rivalry, happens largely unconsciously, and it wastes the cognitive resources of billions of people the world over.

The ongoing fight for status has other downsides. While competition can make people focus, there will always be losers in a status war. It's a zero sum game. If everyone is fighting for high status, they are likely to feel competitive, to see the other people as a threat. So fighting for status can impact relatedness, which means people won't collaborate well. Clearly it would be useful to reduce status threats in the workplace.

Emily tried one possible strategy for doing so during the call with Colin: she tried to put down her own status when she saw that Colin was threatened, saying, "It's been a tough day for me, too, you know. I am not doing so great in my first few weeks as a boss here." Many people do this type of "leveling" intuitively, without knowing why. If you want to have a potentially threatening conversation with some- one, try talking down your own performance to help put the other person at ease. It didn't work for Emily with Colin, but it can some- times be helpful. Being brought down off a pedestal in someone's mind may reduce his threat level.

Another strategy for managing status is to help someone else feel that her status has gone up. Giving people positive feedback, point- ing out what they do well, gives others a sense of increasing status, especially when done publicly. The trouble is, unless you have a strong director, giving other people positive feedback may feel like a threat, because of a sense of a relative change in status. This may explain why, despite employees universally asking for more positive feedback, employers seem to prefer the safer "deficit model" of management, of pointing out people's faults, problems, and performance gaps, over a strengths-based approach.

These two strategies—putting your status down and others' status up—only help other people with their status, and may actually threaten yours. So where can you get a nice burst of confidence-inducing, intelligence-boosting, performance-raising status around here, without harming children, animals, work colleagues, or yourself?

There's only one good (non-pharmaceutical) answer that I've found so far. It involves the idea of "playing against yourself." Why does im- proving your golf handicap, your likes on Instagram, or your position in World of Warcraft feel so good? In big part, it is because you raise your status against someone else, someone you know well. That some- one is your former self. "Your sense of self comes online around the same time in life when you have a sense of others. They are two sides of same coin," Marco Iacoboni explains. Thinking about yourself and about others uses the same circuits. You can harness the power of the thrill of "beating the other guy" by making that other guy (or girl) you, without hurting anyone in the process.

Think of Emily and her new team, who are already uncomfortable about Emily, a previously equal-status person, now being their boss. If Emily played the status card at all, trying to be better than her team, it would come off badly. But if she worked hard to improve herself, focusing on her own skill sets, without trying to better her peers, she might be less of a threat. To play against yourself gives you the chance to feel ever-increasing status, without threatening others. And if you share your progress (and challenges) with others, it can increase a sense of relatedness, too. I have a hunch that many successful people have worked all this out and play against themselves a lot.

To play against yourself you have to know yourself. And this requires a strong director, but it also builds a stronger director as you focus on how you are growing. And here's a really big idea: one way you might play against yourself could be to work on improving your capacity to catch your brain in action. You could practice getting faster at things such as labeling and reappraising, reading other people's states, or developing a quiet mind when needed. As you improve your skills in this area, you raise your own status, without risking other people's status. You increase relatedness if you share what you notice; you even build your director. And of course you make better decisions, deal better under pressure, and collaborate better with others.

PUTTING ON A SCARF

By now you have probably noticed that many of the primary rewards and threats discussed in the last few scenes have features in common and are interconnected in a number of ways. For example, on the ill-fated conference call, Colin experienced more than just a status threat. He also experienced uncertainty, a sense of decreasing autonomy, and a feeling of inequity.

I noticed a surprising pattern in the several years of putting this book together. I saw that there are five domains of social experience that your brain treats the same as survival issues. These domains form

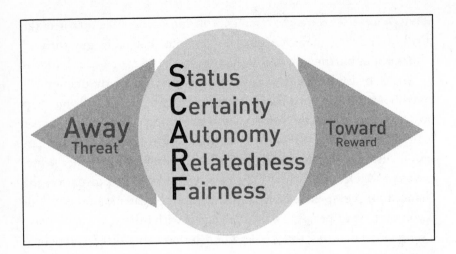

a model, which I call the SCARF® model, which stands for Status, Certainty, Autonomy, Relatedness, and Fairness. The model describes the interpersonal primary rewards or threats that are important to the brain. Getting to know these five elements strengthens your director. It's a way of developing language for experiences that may be otherwise unconscious, so that you can catch these experiences occurring in real time, or even better, avoid strong threats ahead of time through better prediction.

Like Colin's challenge during the call with Emily, some of life's most intense emotional reactions involve a confluence of the elements of SCARF. Imagine your status has been attacked, publicly, unfairly, in a way that you don't understand and can't do anything about. People who experience events like this (say, being unfairly treated at work or being attacked falsely in the media by someone you thought was a friend) find that the pain from these events can take years to recover from. A study of social pain in 2008 found that social pain comes back when you think about it again, whereas physical pain doesn't. Giving someone a bruise on the arm as punishment for a mistake may, in theory at least, be a "kinder" punishment than attacking their ideas in public. (I am not condoning physical violence in any sense, of course. Just making a point.)

On the plus side, if you can find ways to increase several of the elements of SCARF at the same time, either in yourself or in others, you have a powerful tool not just for feeling great but for improving performance, too. Think about what it feels like when you interact with someone who makes you notice what's good about yourself (raising your status), who is clear with his expectations of you (increasing certainty), who lets you make decisions (increasing autonomy), who connects with you on a human level (increasing relatedness), and who treats you fairly. You feel calmer, happier, more confident, more connected, and smarter. You are able to process richer streams of information about the world, which feels like the world has gotten bigger. Because this experience feels so good, you want to spend time with this person and help them any way you can.

While all the elements of SCARF are important, in this scene it's mostly status that has thrown Emily's plans into disarray. Let's find out how differently things might have gone if she had understood the deep human need to protect one's sense of status.

THE BATTLE FOR STATUS, TAKE TWO

It's 4:00 p.m. The conference call ended over an hour ago, the team in disarray. Emily tries to work on something else, but her mind swarms with unanswered questions from the call. She wants to make sense out of the situation, as the brain likes to do every moment, but all she sees are impasses. She deletes and files emails for a few minutes before her director kicks in, recognizing she is avoiding a call that needs to be made.

She dials Colin. A quiet internal voice, a distant connection, tells her to wait and prepare, but the signal is fleeting. She's mad as hell at Colin for upsetting Leesa. Emily's director kicks in again, telling her to stop and reflect, not to rush into this situation without at least tapping into this quiet voice. She is riled up, and she knows she needs some help to balance out her state, or she will make mistakes. She stops dialing Colin's number and instead calls Paul at work to ask for his help. She tells him she is having a tough afternoon. Labeling the situation helps a little. She asks him to tell her about the kids, to change her focus so she can perhaps enter a positive state of mind.

As Paul describes the bonding time he has been having with Michelle, Emily visualizes her oxytocin kicking in and her cortisol level going down. Whether it's possible to feel your neurotransmitters increase has not yet been studied, and is not the point: focusing on an expectation of increased calm will increase your calm; that's the power of expectations. In a calmer and happier state after just a few minutes of talking about the kids, Emily has an insight. She realizes that Colin's status was threatened, and that both he and Leesa are locked in a battle to be "right" about this event. She forms a plan, thanks Paul, and calls Colin.

"I thought that would be you," Colin says.

"Colin, I made a terrible mistake. I didn't manage that conference call well, and the result was that you had to endure a feeling of being put down in front of your peers. It must have been awful. I am very sorry. It's my first two weeks in the job, and I am not on top of things yet."

Colin is taken aback. He was ready for battle; this wasn't at all what he was expecting from the call. He exhales a few times to try to change his state. He hadn't put words to what he was feeling, but hearing this statement from Emily relieves his anxiety. He now knows why the call was so enraging to him.

Emily continues, sensing a little more is needed to bring Colin back from the edge. "Colin, it's my fault for not setting up the call properly. I should have set it up so that everyone got to connect, instead of putting you on the spot like that."

It would be unfair for Colin to be harsh to Emily now. She has allowed herself to be wrong, which has raised his status by dropping hers. While this might look "unfair" to Emily, her own sense of status is okay as she chose to take this route; she is in control. And her job is to build a team, to get the best performance out of others, not to worry about her own feelings.

"Yeah, I guess I will have to forgive you," Colin says with a joking tone. Both of them are relieved and let out little sighs as the tension goes down.

Colin had been dreading the expected fallout from the conference call. The strong away state created by this negative expectation has been displaced by an unexpected reward, in the form of his status going back up. This interaction brings a rush of positive dopamine, oxytocin, and serotonin. Emily and Colin now feel connected, and stay on the line to talk about the team and their upcoming projects.

In the midst of talking about another project, Colin agrees to call Leesa and apologize. He realizes that his remarks were not appropriate, even though his usual jokes would have been fine face-to-face. It wasn't so hard to see this when he was no longer fending off a threat to his status.

Thirty minutes later, Leesa calls Emily. They patch up their relationship and plan some other events. Emily is pleased she understood the brain's deep need to manage a sense of status in this tricky social situation. Things could have gone very differently. But now it's time to head home to be with her family.

Surprises About the Brain

- Status is a significant driver of behavior at work and across life experiences.
- A sense of status going up, even in a small way, activates your reward circuits.
- A sense of status going down activates your threat circuitry.
- Just speaking to your boss or a person of higher status generally activates a status threat.
- People pay a lot of attention to protecting and building their status, probably more to this than any other element of the SCARF model, at least in organizations.
- There is no one fixed status scale; there are virtually infinite ways of feeling better than others.
- When everyone is trying to be higher status than others there is a decrease in relatedness.
- Because we perceive ourselves using the same circuits we use when perceiving others, you can trick your brain into a status reward by playing against yourself.
- Playing against yourself increases your status without threatening others.
- Status is one of five major social domains that are all either primary rewards or threats, which form the SCARF model for Status, Certainty, Autonomy, Relatedness, and Fairness.

Some Things to Try

- Watch out for people's status being threatened.
- Reduce status threats in others by lowering your status through sharing your own humanity or mistakes.
- Reduce status threats in others by giving people positive feedback.
- Find ways to play against yourself, and pay a lot of attention to any incremental improvements. The slightest feeling of improvement can generate a pleasant and helpful reward.
- Playing against yourself to improve your understanding of your own brain can be a powerful way of increasing your performance.

ACT IV

• • • • • • • •

Facilitate
Change

C hange is hard, and facilitating change in others is even harder. Research indicates that we have more influence over other people, but less control, than previously thought. In this final act, the story takes a slight turn, focusing less on knowing your own brain and more on how to apply what you know to changing others, first one-on-one and then in a group.

The brain constantly changes based on external factors, but it can also be changed by shifting people's attention. Shifting other people's attention from a threat state to focusing on what you want them to focus on is the central challenge to creating real change.

In act 4, Paul learns why getting another person to do what he wants can be so tough, and discovers a faster and easier way to increase someone's performance. Later Emily and Paul get home and discover the difficulty of changing the way a group interacts, and learn a new approach to creating change within a wider culture.

SCENE 13

• • • • • • • • •

When Other People Lose the Plot

t's 4:30 p.m. Paul gets an email from Eric, the supplier collaborating with him on the school project. Eric has written to say that the project is behind schedule and the school principal is upset. Paul goes to write an email back and then decides to call instead, remembering the lesson he learned from earlier today with Ned.

Eric is defensive the moment he answers the phone. It's only his second project with Paul, and he wants to look good. He explains that the project is over budget, four weeks over deadline, and still not finished, but it's all because of annoying changes the client keeps making. Between the threat to his status and his uncertainty as to what Paul might say, Eric's limbic system is in overdrive.

Paul isn't his best self, either. His reputation in a community of hundreds of parents is at risk, generating a deep status threat. And when Paul thinks about facing up to the principal, a decades-old limbic memory of being in trouble and going to see the principal resurfaces. Paul has the urge to start the call by yelling at Eric, but he knows anger will only make things worse.

"So, why did it go wrong? What was the problem here?" Paul asks, trying to hold back his emotions.

"Look, it's not my fault," Eric replies. "The client keeps changing the brief midstream, and each time they do, we end up doing more work. I can't help it if they don't know what they want."

"Here's the thing, Eric . . ." Paul pauses, thinking about how he can best give Eric some feedback. He remembers something from a book about a "feedback sandwich," and tries starting with a positive item, softening the feedback a little. "Eric, you did great work on the first project we worked on, but this one is in a bit of a mess. I'm sure you will do good work again, but there's a real problem here—"

Eric interrupts him. "Are you trying to say it's my fault? You know the client changed the brief. You were there when it happened." His voice is rising in anger. Despite a few positive words from Paul, Eric's limbic system is primed to fight. As well as a status threat, Eric also feels Paul is being unfair.

Paul feels his own anger welling up. He probably wouldn't have given Eric feedback if he had been thinking straight, and now it's made things even worse. This is a turning point in the conversation: if Paul allows himself to get emotional now, a long argument will erupt, the third tiff in the few months they have been working together. He pauses a moment, letting his director observe the scene to find other paths to take. With a lot of effort, Paul manages to reappraise, focusing on the fact that Eric is still a new consultant making easy mistakes that many others would. He will still be a good collaborator in time. This reappraisal helps Paul's anger subside. Eric's mirror neurons pick up the emotional change, and he start to feel calmer now, too.

Paul thinks about different approaches he can take here. Giving direct feedback isn't working. He decides to try to be more helpful and work with Eric on the cause of the problem.

"Look"—Paul speaks more slowly to calm Eric down—"I am not here to give you a hard time. I'm sure you did your best."

"I appreciate that. Thanks." Eric comes back from the edge a little more.

"Let's just talk about this logically and break it all down, step by step," Paul continues. "Why do you think this went wrong?"

Eric explains the details of what happened over the last few weeks, culminating in the angry call today from the school principal. For more than forty-five minutes they discuss the project from every possible angle. It feels like wading through mud, but Paul can't think of another way to find the source of the problem. Finally, after going over the same issue four times, they decide to agree that this was just a "normal" problem with a new client. While this "solution," a type of reappraisal, helps put the issue to one side, it doesn't address what to do with the principal now. Paul gets impatient and decides to put forward a solution. He suggests that Eric call the principal back and go through the original brief. Eric rejects the idea, and another debate starts. Twenty minutes later, Eric agrees to think about the problem further on his own.

Paul thinks he knows what the answer is: to write up a new contract with the client. If only he could convince Eric of his perspective. The conversation has now taken more than an hour when it should have taken only ten minutes. Paul wonders if working with others is worth the pain and effort.

This complex situation in a nutshell: The school software project Paul and Eric are working on has gone off the rails. Paul wants to help Eric sort it out. Eric is stuck at an impasse, and Paul is angry because of his own sense of threat. Paul then tries a textbook feedback technique. It's the wrong strategy; especially for someone already feeling threatened. Then Paul tries a more "rational" approach and attempts to break the problem down. The two men find themselves lost in the details, going in circles. Paul proposes a solution, which Eric rejects without a second thought.

After giving up on providing feedback, Paul is taking a logical approach to helping someone else solve a problem. He tries to understand the source of Eric's problem, and then makes suggestions. I call this approach the *default* approach to helping people. What Paul doesn't realize is that this default approach is inefficient for solving human problems, and even has some undesirable side effects. While Paul is great at finding software glitches, he needs to change his brain to be better at increasing other people's performance.

THE TROUBLE WITH FEEDBACK

Giving others feedback is often the first strategy people use to facilitate change. Yet, surprisingly, giving feedback is rarely the right way to create real change. While there are many "techniques" to improve the performance of feedback, people miss the basic reality of this approach: feedback creates a strong threat for people in most situations. The statement "let me tell you what others have been saying about you" is one of the fastest, easiest, and most consistent ways of making someone deeply anxious.

Paul's first attempt to help Eric involved a "polite" approach to giving feedback: he said something nice, then attacked Eric's status, and then said something nice again. To me this is like an "arsenic sandwich": the bread might make the meal appear more palatable, but it's still going to kill you.

Feedback is something that organizations worldwide have mandated over the last decade in the form of the annual "performance review." Mike Morrison, at the time the dean of Toyota University in Los Angeles, commented that annual performance reviews, "Essentially just reduce performance for six days each year: three days while people prepare for it, and three days recovering from it." Performance review training manuals tell managers to give "constructive performance feedback." The problem with "constructive performance feedback" is that, like a wolf sniffing a meal across a field, even a subtle status threat is picked up unconsciously by our deeply social brain, no matter how nicely it's couched. As "constructive" as you try to make it, feedback packs a punch. The result is that most feedback conversations revolve around people defending themselves. There has to be a better way to facilitate change in others.

THE PROBLEM WITH PROBLEMS

When feedback didn't work, Paul's "better way" was to drill down to find the cause of the problem. He wanted to be rational. This deductive approach to problem-solving works in many domains of life, such as finding out why your car has overheated or your software malfunctioned. Cars and software are linear systems. Problems at work, like corporations and people generally, are often complex and dynamic.

Imagine you are in a new city and need to be at the airport by 2:00 p.m. to fly to a client meeting elsewhere. You plan to take a taxi to the airport, but are not sure when to leave your hotel. In this case you hold three ideas on your stage at once: the idea of "being there at 2:00 p.m.," the idea of "leaving the city," and the idea of "taking a taxi." In a sense, you are creating a gap between these three ideas, then seeing what information emerges to fill this gap. Let's say the answer that comes to you is "Leave at 1:00 p.m." You've used deductive reasoning, the human default approach that's great for linear situations, to solve an external problem. So far, so good.

Now it's 1:00 p.m., and you're trying to hail a taxi, when it starts to rain. Ten minutes later, there's still no taxi. You start to panic about missing your flight. It's too late now to get a bus or train. You start to get annoyed with yourself, and begin to hold three new questions on your stage: Why didn't you think to check the weather? Why didn't you think to ask someone about going to the airport? Why are you so disorganized? You try to close the gaps between these questions, to find the information that completes this circuit. As you do so, your medial prefrontal cortex is activated as you scan back through memories in your hippocampus. Focusing your attention inwardly in this way causes you to remember several recent stressful situations, bringing the related stress back to mind. The questions you asked yourself changed the state of your system. You decide that the problem is you have been more stressed lately. To the outside world, it looks like you are daydreaming. A taxi stops a few feet away and someone else jumps in, straight out of a store, hardly a drop of rain on him. As you yell out at the driver, another vacant taxi swerves to avoid picking up the crazy person yelling at cars in the rain. In your worked-up state you call the client to cancel the meeting, complaining about the city traffic that made you miss your flight. The client isn't impressed.

In this story, applying a deductive, problem-solving approach to yourself had unintended consequences. Bringing to mind problems . . . well, brings them to mind. Unless you take care to label your emotions just at a high level, and not dwell on them, bringing problems to mind will increase limbic arousal, making it harder to solve them. "I'm feeling like an eight out of ten in my anxiety level" gets a really different outcome than "This rain is driving me crazy and I am a mess and nothing is going right today." Solving difficult problems, after all, involves getting around an impasse. This requires a quiet and generally positive and open mind, as we learned in scene 6. Getting lost in large amounts of history and detail doesn't make your brain quiet at all.

In the taxi story, while you did make connections, these connections didn't help you get to the airport. A similar thing happened with Paul and Eric as they drilled down into the details of the project. They solved problems, but the problems they solved weren't helpful to their real goal. This is one of the traps of problem-solving: solving any problem creates a little rush of dopamine, which drags you farther into the story. The key is to make sure you solve the right problem, which means the most *useful* problem, not just the most *interesting* one.

When you follow a thread down to the root of a problem, as interesting as it might seem at first, you often end up arriving at the conclusion that there is "too much work," "not enough money," or "no time." Paul and Eric hit a dead end like this, with the idea that the problem was just a "new client" issue, just something that "happens" sometime with new clients. These types of answers are rarely helpful, and worse still, they leave you exhausted, due to the downward spiral created. The more negative connections you make, the less dopamine you have, the fewer resources you have for solving the next problem, and the more negative connections you make. And on it goes. In this low-energy state, everything looks hard. Increasingly risk averse, you don't have the motivation to take action. Eventually, all you feel like doing is taking a nap. What's needed here is a strong director to catch the wrong train of thought early enough before the downward spiral takes over.

If having a problem focus is so unproductive, why do people do it so often? One answer is that the problem-focused approach looks "safer." Remember that the brain dislikes uncertainty. The past has lots of certainty; the future, little. Going into the past might make you want to take a nap, but finding answers amid uncertainty can feel like diving into a deep and unknown ocean.

There's another reason that focusing on problems is so ubiquitous. When you ask yourself or another person a question, where does the information come from to close the gap the question creates? From the billions of circuits in your brain that represent memories of the past. If you don't look to the past, where will you find circuits to connect to? The brain has few circuits for the future. Conceptually speaking, electrical impulses are more likely to travel along existing paths, because this requires less energy than traveling along paths that don't yet exist.

FINDING THE SOLUTION

Going back to the imaginary challenge about getting to the airport. Once the rain starts, an alternative path would have been to solve a different problem, perhaps this one: "It's raining. There are no taxis. Where will I find one?" This question would focus you on the external world instead of the internal one. Focused on the external, you would notice lots of occupied taxis and realize that you are close to a subway stop, where taxis might drop people off. When you see a taxi in the distance heading toward the subway, you're the first person to step off the curb toward it. That's because your neurons relevant to "taxi heading toward the subway" have lit up in anticipation of seeing this event, making you the first to notice subtle signals, even when those signals are only tiny changes in the pattern of light entering your eye from a taxi changing lanes three hundred feet away in the rain.

The difference between the two taxi scenarios is based on one key decision: to focus on the desired outcome (getting a taxi) rather than on the past. Attention went to your goal, rather than to your problem.

The decision to focus on an outcome instead of a problem impacts brain functioning in several ways. First, when you focus on an outcome, you prime the brain to perceive information relevant to that outcome (find a taxi), rather than to notice information about the problem (not getting to the airport). You can't be looking for solutions and problems at the same time. That would be like trying to hold two large numbers in mind at once, and trying both to add them up and to multiply them at the same moment. Your actors can play only one scene at a time. And if it's a solution that's needed, it's more useful to prime your brain to notice information relevant to the solution itself.

When you look for solutions, you scan your environment widely for cues, which activates more of the right hemisphere of the brain, rather than drilling down into information that activates the left hemisphere. Activating the right hemisphere is helpful for having insights, which is how complex problems are often solved.

When you focus on problems you are more likely to activate the emotions connected with those problems, which will create greater noise in the brain. This inhibits insight. Whereas focusing on solutions generates a toward state, because you desire something. You are seeking, not avoiding. This increases dopamine levels, which is useful for insight. And if you are expecting you might find a solution, these positive expectations help release even more dopamine.

In all these ways, focusing on solutions can significantly increase the likelihood of having insights, and even make you feel happier. However, focusing on solutions is not the natural tendency of the brain. Solutions are generally untested, and thus uncertain. It takes effort to dampen down the threat created that comes with uncertainty. To focus on solutions, sometimes you need to activate your director, veto your attention going to problems, and gently nudge your brain in a direction it would rather not go. Thus, people without a strong director (or those whose threat response has pushed their director aside) will naturally tend to focus more on problems.

THE DOWNSIDE TO SUGGESTIONS

There is another more subtle challenge with focusing on solutions. Because problem-solving can be exhausting, it's logical to want to conserve energy and head straight to solutions. The difficulty with this strategy is that when trying to help someone else solve a problem, people often end up simply providing a set of solutions to the other person.

This is what happened with Paul. He went ahead and suggested a possible solution to the school principal issue, which Eric then discounted out of hand. The source of the difficulty here lies in who comes up with the solution. Paul's suggestion makes him look smarter, and Eric less smart. This impacts their relative status, which Eric is likely to fight against. The better Paul's answer is, the more likely Eric might resist it. It's bizarre. (The exception to this, of course, is for things such as finding a password or a basic piece of information when it might insult someone's intelligence to ask a question.) Paul's giving out suggestions also threatens Eric's autonomy: it's no longer Eric's choice to follow a specific path.

If Eric had come up with a solution on his own, his status would have gone up, along with his sense of autonomy and even some certainty. He would also have gotten a nice buzz from the energy of a novel insight forming in his own mind. The "aha" experience is so much more energizing than the "a-duh" experience. The positive buzz from an insight might have helped Eric push past the uncertainty implicit in doing something differently.

Despite the inefficiency of giving advice, people rush to dish out solutions because waiting for someone to come up with their own ideas requires effort. First you have to hold back your desire to solve the problem yourself, which takes inhibition, an energy-hungry process. It can feel like staring at someone trying to solve a crossword puzzle clue you know the answer to—a little painful! You also have to work hard to dampen your arousal from the uncertainty of what solution the other person will come to, the lack of autonomy you might now experience because someone else is making the choices, and the possible threat to your status if the other person comes up with a good idea you didn't have.

There is a big irony here. Because it looks like so much effort to help other people solve problems, smart business leaders the world over spend millions of hours thinking hard about other people's problems, yet the harder these leaders think, the more others feel threatened, and discount the suggestions made. There has to be a better way.

FROM CONSTRUCTIVE PERFORMANCE FEEDBACK(CPF) TO FACILITATING POSITIVE CHANGE(FPC)

The clue to the better way here lies in Eric's own response at the end of this scene: he wants to go away and think about the issue. Eric isn't going to take action until he has an idea that fits with his own thinking. In his current over-aroused state, he quickly rejects external ideas. Given that Eric is at an impasse, Paul needs to help him find an insight to solve this problem. If Paul can't make direct suggestions, why can't he just give Eric some clues about what to think about, perhaps posing a good suggestion as a question?

You met Dr. Stellan Ohlsson back in scene 6. He is the scientist in Chicago who studies impasses. In one study, Ohlsson set up situations where people had impasses and then tried two techniques: giving other people clues about what *not* to think about, and giving clues about what they *should* think about. "The effects of these strategies are negligible," Ohlsson explains. Ohlsson finds that when someone hits an impasse, telling her what not to think tends to help only 5 percent of the time. Giving people clues about what they should think about tends to help only 8 percent of the time. One of the most common strategies human beings use to help one another solve problems involves these techniques: giving advice about what to do or what not to do. Ohlsson shows that this is only marginally effective. The other strategy people use is to dive into the problem. These two approaches make up the majority of the human default approach to helping others get unstuck. Clearly the intuitive human response to helping others is far from efficient and needs rethinking.

What can Paul do here? As you learned in scene 6, people have insights when their brain is in a specific state. Insights happen when people think globally and widely rather than focusing on the details. Insights require a quiet brain, meaning there is an overall low level of electrical activity, which helps people notice subtle internal signals. As people are often already anxious when stuck at an impasse, and anxiety generally makes people's views narrow and their brains noisier, it's important to reduce people's anxiety and increase their positive emotions—in other words, to shift them from an away state to a toward state. A great way to do this is using elements of the SCARF model.

You could help the person increase her sense of *status*, perhaps by encouraging her. Or increase someone's sense of *certainty* by making implicit issues more explicit, say, by clarifying your objectives. Or increase a person's sense of *autonomy* by ensuring that he is making the decisions and coming up with the ideas, not just listening to your suggestions.

Another useful step is to help people simplify a problem into as few words as possible, to reduce the load on their prefrontal cortex and thus reduce its overall activation level. Sometimes reducing a problem to one short sentence can be enough to bring about insight on its own.

Once the other person is in the right state of mind, and you have a problem stated simply, your job, according to the research, is to help people reflect, though in a quiet way. You want people to look inward, but without dwelling on the details of the problem. This is a subtle technique, but once you see it a few times, it soon becomes clear. Your goal is to facilitate the state of mind that you have when you first wake up, when you easily connect distant ideas, and subtle thoughts can rise to the surface.

The questions to ask at this point should focus people's attention on their own mental process, at a high level. As Mark Beeman says in the first edition of the *NeuroLeadership Journal*, you can increase the likelihood of insight through "variables that increase attention to subtle connections." You want people to focus on their own subtle connections, and a simple way to do this is just to ask about subtle connections.

Paul could have asked Eric questions such as:

If you stop and think more deeply here, do you think you know
   what you need to do to resolve this?
What quiet hunches do you have about a solution, deeper inside?
How close to a solution are you?
Which pathway to a solution would be best to follow here?

I give many more examples and background to this approach in my last book, *Quiet Leadership*, but the principle is simple: Help the other person notice the subtle, high-level connections in his own thinking, which will make insight more likely. While you can't control insight, you can influence it more than people realize. What you are doing is facilitating in others the ARIA model (Awareness, Reflection, Insight, and Action), which I introduced back in scene 6, as a fast way of getting past impasses.

One big advantage of this technique is that it raises people's status by implicitly saying, "You have good ideas. Let's explore what your good ideas are, rather than think about mine." When you ask people to pay attention to their own subtle internal ideas, you are also activating their director. This will also help dampen their overall arousal.

These kinds of questions generate a whole new thread to follow. Instead of *your* looking for a gap in the form of the source of another person's *problem*, the other person is finding a gap in his own *thinking process*. It's not *you* searching for *problems*; it's *him* searching for gaps in his *thinking process*. You want people to look for assumptions or decisions that don't make sense upon further reflection.

This approach is so different from what normally happens in the workplace. The poor quality of feedback is one of the biggest complaints by employees everywhere. This is an unfortunate cycle that new managers often go through: To begin with, they give lots of feedback, thinking people will appreciate this. Then they discover how people are easily threatened by feedback. They notice the long arguments and wasting time, and soon learn to not give feedback, but to avoid it. Then, at some point, they are forced to give feedback—by a performance review, or a mandate from their own boss. So their next technique is to waffle—to not say much at all—to avoid threatening the other person. One study showed that feedback does nothing or makes things worse more often than it is helpful. The brain research explains not just why this cycle happens, but also a new approach that is likely to work better.

For Paul to take this new approach, he will have to activate his director to inhibit his attention going into the problem or straight to giving out a solution. If you don't practice vetoing your desire to solve other people's problems, your default approach, it's easy to waste time in unnecessary discussions driven by people protecting their status. When your objective is helping other people be effective, sometimes to move fastest you have to put on your own brakes.

THE SIGNIFICANCE OF STATUS

This idea of letting other people find their own solutions isn't just relevant to managing projects. Tremendous resources are wasted by people trying to protect their status in all sorts of situations. "One out of fifty people in college are good writers, in my experience," Lieberman explains. "I make a point that I do not grade my students on the draft they have written of their paper. I grade them on how successfully they have critiqued their work. I build an incentive structure for being able to successfully attack their own work. The better they do that, the better they do in class."

When you review your own work, there's an incentive to convince yourself that the work is good. You don't want to look bad to another person. Eric, for example, is convinced he's done nothing wrong with the school project, especially when Paul thinks he might have. When Eric looks into his own thinking, with the idea of protecting his status in mind, all he sees are all the things he has done right. His brain is primed only to notice what he has done right.

Lieberman plays with this traditional incentive structure. He grades his students' work based on how well they incorporate their own earlier criticisms into their writing, on how much they improve. He links people's sense of status to how much they can change. Their status is linked to the criticizing side, instead of to being criticized. It's like being a masochist—you feel good about (metaphorically) beating yourself up. Lieberman explains the dramatic impact this has. "My students say, 'I read my paper with completely different eyes. I finally could read my own paper as if it was someone else's. You could see all the errors glaring at you.'" When you read anyone else's writing, all the mistakes are obvious. In your own work this is normally much harder to see. This probably also explains why the act of writing itself is easier when you leave a gap between writing and editing: you've forgotten it's you who wrote the words. You can see your sloppy sentences with the eyes of a stranger, someone with no agenda to protect poor quality.

Lieberman has shown that people are, in theory, capable of giving themselves feedback, especially if their status isn't threatened. They may even be more capable if their status is harnessed. But it's not status itself that is the active ingredient in the change process; Lieberman gets people to activate their director, using status as the reward to do so.

The more you can help people find their own insights, the easier it will be to help others be effective, even when someone has lost the plot on an important project. Bringing other people to insight means letting go of "constructive performance feedback," and replacing it with "facilitating positive change." Instead of thinking about people's problems and giving feedback or making suggestions, change can be facilitated faster in many instances if you think about people's thinking, and help others think about their own thinking better. However, letting go of the default approach to problem-solving requires working against the way your brain wants to go, which requires a good director. And to be most effective at bringing others to insight, your goal is to activate other people's director, too.

With all this in mind, the question here now is what would Paul have done differently, and how would it have turned out, if he understood and applied all the ideas in this chapter. Let's find out.

WHEN OTHER PEOPLE LOSE THE PLOT, TAKE TWO

It's 4:30 p.m., and Paul gets an email from Eric. The school project has gone off the rails. Paul goes to write an email, then decides to telephone Eric instead. Eric is defensive from the first word, his status threatened. With so much at stake, Paul's immediate response is to get angry, though he manages to suppress this response.

"So, why did it go wrong, what was the problem?" Paul asks. Just as he says this, he remembers noticing a pattern in similar situations: focusing on a solution often brings better results than focusing on a problem. He turns the question around. "You know what. Don't worry about what the problem is. It's not that useful. I'm sure you did your best. Let's think about what we can both do here to rescue the situation. I'm not going to give you a hard time. Let's work together on this, okay?"

Eric sighs. He was expecting to have to defend himself, but Paul's positive approach has disarmed him. However, he is still overaroused and not thinking clearly. "I have no idea what to do," he says. "All I can think of is the client making all these changes." The problem as Eric sees it has become deeply primed, inhibiting other ways of thinking.

Paul thinks he has seen this type of situation before and jumps to a solution.

"Why don't you go back and ask the client to rework the contract? That's what I would do in this situation," Paul says.

"I can't do that," Eric responds.

"Why not?"

"You don't understand. It's a big project, and the person I am dealing with is really annoyed."

Eric is defending himself again. Paul stops and reflects for a moment and realizes he is accidentally making the connections, instead of helping Eric make them. He needs to step back and help Eric do the thinking.

"Can I ask you a few questions, to see if I can help you solve this?" Paul says.

"Sure," Eric responds. Asking someone else for permission to stretch their thinking can create a nice positive flush from a sense of increased status and autonomy.

Paul hesitates for a moment, and vetoes several paths his attention wants to go in: he would much rather make a suggestion or focus on the problem. Then he clicks on something and launches in.

"Tell me what your goal is here, in one sentence."

Eric reflects for a moment, activating just the right circuitry for insight. Then he sees something. On the other end of the phone, Eric's eyes flicker as a new connection forms.

"I think the central challenge here is knowing how to get the principal to be happy again."

"How many different strategies have you tried so far, to solve this problem?"

Eric is thrown by the question. It makes him think. He responds after a moment of reflection. "Well, I haven't really tried any yet. But I have a few ideas, maybe three or four. They all seem to be along a similar line I guess." Eric's eyes go up as he observes his own thinking process. He is looking into his own thinking, not at the details of the project, but at the connections holding this impasse together. His right hemisphere is coming alive.

"What other directions do you think might be worth trying?" Paul asks.

"I don't know. I guess the principal is really angry his expectations weren't met. There's nothing we can do now, except . . ." In that moment, Eric has a central insight. He sees things in a whole different way. The energy released by this insight creates a positive mental state, like a storm clearing in his mind.

"Maybe I need to go back to the brief and create a new set of expectations," he continues. "Maybe there's no answer except this one: perhaps we didn't take enough care with the contract." He lets out a sigh. To have this insight means potentially being in the "wrong"—something difficult to admit when one is feeling a strong level of threat. With this insight in mind, Eric has already decided what to do, and Paul can relax. His hard work is done, in less than ten minutes. Eric is on track with what he needs to do, the project should get back on the rails now, and Paul didn't have to fight like he has had to in the past. With the extra time and positive mental state from mirroring Eric, Paul finds himself thinking about his plans for tomorrow and how he can organize his day best. Before long he hears the garage door open, and it's time to get the family together.

In summary, trying to change other people's thinking appears to be one of the hardest tasks in the world. While the easy answer may seem to be to give people feedback, real change happens when people see things they have not seen before. The best way to help someone see something new is to help quiet her mind so that she can have a moment of insight. As you have insights, you change your brain, and by changing your brain you change your whole world.

Surprises About the Brain

- Giving feedback often creates an intense threat response that doesn't help people improve performance.
- The problem-solving approach may not be the most effective pathway to solutions.
- Providing suggestions often results in a lot of wasted time.

- Bringing people to their own insights is a fast way of getting people back on track.

Some Things to Try

- Catch yourself when you go to give feedback, problem solve, or provide solutions.
- Help people think about their own thinking by focusing them on their own subtle internal thoughts, without getting into too much detail.
- Find ways to make it valuable for people to give themselves feedback; reward them for activating their director.

SCENE 14

• • • • • • • •

The Culture That
Needs to Transform

t's 6:00 p.m. Emily lunges toward the front door, her email inbox overflowing with work to do after dinner. She remembers walking to this same door a few years ago, arriving to the sound of unsteady footsteps racing to greet her. For a moment, as she struggles to open the door, the same positive neurochemistry she felt then courses through her brain.

Emily walks inside to see Michelle on the couch with her headphones on, eyes closed, and head bobbing to 130 beats a minute. Listening to repetitive patterns of noise interspersed with slight variations is mildly pleasing to an adult brain. To a teenage brain easily fired up by small neurochemical shifts, the same patterns can be completely engrossing.

"Hi, Mom," Josh says without looking up from the TV.

Emily's dopamine levels crash as reality sets in, her unconscious expectations shattered.

"Will you all do something useful, please?" she yells out, suddenly turning off the television. She is unable to hold back her increasing arousal on an empty stomach. Josh goes to yell out, then sees the look on his mother's face and decides to stay quiet. Michelle is oblivious to Emily's presence, until she feels her headphones pulled off and sees an angry mother inches away from her nose. The shock of the unexpected change is overwhelming. In a fraction of a second, Michelle's brain manipulates her vocal cords so that the noise made by the exhalation of a breath becomes a word most appropriate to this sudden arousal. Before she even knows she has done it, Michelle has yelled out an expletive that hasn't been heard in this house before.

Emily has been unhappy with how her family has been communicating with one another for a while, but she has been keeping this thought suppressed until now. The swear word from Michelle is the final straw. Tonight, Emily is ready to tackle the situation head on and change how her family communicates.

An hour later, after time apart to cool down, dinner is on the table, home-delivered Chinese food.

"I'd like to have a family conference tonight," Emily announces. Suppressing her emotions for an hour has built them up, and the children sense a problem.

"No way, Mom. We had one last year," complains Josh, trying to be funny. Josh gets a strong threat response talking about feelings. Recently he has been watching horror movies with friends, a modern version of an ancient communal ritual where young adult males practice emotional regulation to prepare for a hunt. Josh now watches scenes he couldn't bear to look at a year ago, but he still can't face emotional conversations with real humans. He clams up and tries to suppress his emotions. Expression seems unmasculine, and reappraisal doesn't feel "real." Josh prefers to be more like his dad, and keep his emotions to himself.

Emily knows this isn't going to be easy, and tries to build a solid case. "Your father and I have talked," she begins, "and we want to make some changes. It's time we thought about how we all get on with each other. It seems like there's no communication here. I'd like to set a goal that we can all work toward."

"Oh, Mom," the two kids say almost in perfect unison.

"I want us to be more of a family, to talk more about what's going on for each of us, and to fight less. Will you take this on as a goal, all of you? I promise we'll take a great vacation this year if we can just get along better and be more of a family."

"Sure, Mom, that's fine," Josh says.

"Sure, whatever," Michelle says, hardly looking up.

Emily feels better for speaking up. This thought had been buzzing around for months, an item in her queue taking up space on her stage, interrupting other thoughts.

Ten minutes later, having hardly spoken a word since Emily's outburst, Michelle and Josh finish their last mouthful of dinner, leap from the table, and head to their rooms to connect with their friends. They yell "bye" from the top of the stairs, without even saying thanks for dinner. The topic of annoying mothers will no doubt feature in their evening text messages and posts.

Emily had a hunch this discussion wasn't going to be the last word necessary to change the kids, but she's still surprised that the conversation hasn't registered at all. It's the third time she has tried to make things different at home, and nothing seems to have worked. She wonders if it's possible to make any change in these kids. She tries to imagine what other incentives might work on them. Or maybe she needs to think about some kind of punishment if they don't change.

Paul and Emily have a long debate over the hour and a half they spend cleaning up the house. They don't get to any solutions, only an increased sense of exhaustion. The only positive is the slight reward they both feel from putting everything in its place, a little dopamine eked out of a subtle increase in certainty. Turning off the lights in the kitchen and yelling goodnight to the kids, Emily heads to her study to catch up on work, while Paul watches a movie.

It's now midnight. Emily checks on the kids, washes up in the bathroom, and collapses into bed, trying not to wake Paul. At last their difficult day is done.

Facilitating change in others isn't easy, as we learned in the last scene. What about facilitating change in several people, or more, at once? Even when you have a deep desire to do so, this seems almost impossible at times.

What Emily and Paul don't know is that their models for creating change need updating. Their attempts to cajole their kids might have worked when they were toddlers, but far more sophisticated techniques are needed now. Emily and Paul want to become better at changing the way people around them interact. They need to change their brains so that they can be more effective at creating change not just in one individual, but in a group of diverse characters. They need to learn how to change a culture.

CHANGE IS HARD

Changing one's own behavior is hard. A study found that only one in nine people who underwent heart surgery were able to change their lifestyle to be more healthy, and these people had the ultimate "motivation"—possible death. Changing other people's behavior is even harder. Changing the behavior of a group of people . . . well, that sometimes appears almost impossible. And while this scene focuses on a situation at home, the ideas in it are relevant across the board, including for all types of work situations.

Part of Emily and Paul's problem is they are using a rather blunt instrument to change behavior, known as the carrot-and-stick approach. It's like trying to fix a watch with a hammer. In this case, Emily offers a holiday to the kids if they communicate better. While nothing gets broken here, nothing changes, either.

The carrot-and-stick approach draws from a field called behaviorism, which emerged in the 1930s. The field built on Igor Pavlov's famous concept of the "conditioned response." Associate the ringing of a bell with food, and a dog will soon learn to salivate at the sound of a bell alone. Many of the behaviorist techniques work well with animals and are still widely used, such as for training police dogs.

The behavioral approach also works well with small children, though using different types of rewards and punishments, of course. One surprisingly effective punishment for kids is a "time out," where a child is put in a corner. From the insights in this book perhaps you can see why this works well: because the child experiences a drop in status and relatedness.

The behaviorists generalized their observations to everyone, and this approach has since become the dominant way of thinking about motivation in society at large. The trouble is, the carrot-and-stick approach doesn't work well with adults. Adults can recognize that someone offering goodies is trying to change them, and they class that person as a threat. Or an adult sensing impending punishment may launch a preemptive strike, insulting his punisher with an attack on her status. Now you have a tit-for-tat war of words, rather than any changed behavior.

So, if behaviorism doesn't work well, why is this model still with us? One reason (other than behaviorism being co-founded by an advertising executive) is the allure of its simplicity. With just two ideas to remember, behaviorism appears irresistibly "certain."

THE POWER IS IN THE FOCUS

We are at the beginning of a new theoretical framework for change that draws from the science of the brain. At the heart of this framework is the idea that it is *attention* itself that changes the brain. It's not the carrot-and-stick approach that creates change, it's what this approach sometimes does, which is to focus people's attention in the right way. How exactly attention changes the brain is still a widely debated topic, however, there are aspects of the science that are mostly uncontroversial, and I focus on these here.

At rest, the brain is noisy and chaotic, like an orchestra warming up, a cacophony of sound. When you pay close attention to something, it's like bringing the orchestra together to play a piece of music. Many neuroscientists now think of attention as being a type of synchrony, of the brain getting in tune and working as a unit. Synchrony is a technical word, which means that different neurons fire in similar ways at the same times.

An orchestra playing together is a good metaphor for attention, as in both cases you have individual units now doing things in synch with other units. When you pay close attention to something, different maps across the brain start working together, copying one another, forming a pattern as a unit. Professor Robert Desimone from MIT studies neural synchrony. He believes that attending to stimuli involves use of nearly all of the brain. A 2006 study by Lawrence Ward at the University of British Columbia and four other scientists found that neural synchrony plays an important role in the integration of functional modules in the brain. They even found that neural synchrony is affected by how noisy the brain is. This links back to all of act 2, the way you can't focus when there's too much neuronal activity, such as over-arousal from sensing a threat.

So, when you pay close attention, many brain regions become connected up in a larger circuit to complete a specific task. As they form this larger circuit, a gamma band electrical wave often occurs in the brain, the fastest possible frequency of electrical activity across the brain. This frequency is thought of in some circles as the "binding" frequency, as it's involved in connecting up disparate brain regions. (This is the same wave that occurs at the moment of insight.)

When different circuits fire synchronously, you invoke Hebb's Law, which says that "Cells that fire together, wire together." Putting all this together, you get an explanation for how paying close attention to an idea, activity, or experience helps create networks in the brain that can stay with you, wired together, sometimes forever.

This idea that attention is the active ingredient that changes the brain is supported by a large body of research called neuroplasticity, the study of how the brain changes. Researchers in the late 1970s originally tried to understand why the brain seemed to change after accidents or illness. This went against existing theories of the brain and was a controversial area of research at the time. Over decades this idea became more accepted in scientific circles, and deeper research has emerged. Studies of stroke patients have since shown that regaining the use of an arm requires focusing attention closely on rehabilitation activities, not just doing movements. Studies with monkeys had similar findings.

A study by research psychiatrist Dr. Jeffrey Schwartz showed that changing the way you pay attention can change the circuitry of the brain not just over months, but even within a few weeks, enough to show up on a brain scan. "The power is in the focus," Jeff would say to me over and again in our meetings. Schwartz, working with Henry P. Stapp, a renowned quantum physicist, and neuroscientist Mario Beauregard, has taken steps to explain the physics of *how* cells that fire together wire together, in a paper called "Quantum Physics in Neuroscience and Psychology." "The act of observing, in and of itself, makes a difference, in the material world," Schwartz explains.

Professor Norman Doidge, who wrote the bestselling *The Brain That Changes Itself*, believes neuroplasticity can occur at an even shorter time scale. At the NeuroLeadership Summit in Sydney, Australia, in 2008, Doidge explained how putting a blindfold on someone creates change in his auditory cortex in a matter of minutes. The change occurs because attention is forced there. It seems that attention can quickly change the brain, if enough attention is paid to stimuli. It's just that attention doesn't tend to go easily to one place and stay there. Learning a new language, for example, is relatively easy; it's just that you have to stop paying attention to your current language to create the new circuits. That's why moving to France is the fastest way of learning to speak French—your attention is forced there all day long.

The brain is mutable. It changes all the time, a disconcerting amount, in fact. It changes based on the lighting around you, the weather, what you eat, whom you talk to, the way you sit, even what you wear. The consistency of the brain is like custard, and its makeup is more like a forest than a computer, always alive, rustling, changing. One study showed that you probably don't even use the same neurons to lift your finger now as you did two weeks ago. The brain is happy to change; it's a happy-go-lucky free agent. It is attention that is the grumpy curmudgeon.

It's not hard to change your brain. You just need to put in enough effort to focus your attention in new ways. Your brain changes in a wide-scale way when you make life choices, such as the choice to learn the piano when you are young. Here you have systems that keep your attention focused, such as music exams to pass to impress your friends. However, as Doidge and others point out, your brain can also change in much more subtle ways, in far less time, even moment to moment.

When you change your attention you are, according to Schwartz, facilitating "self-directed neuroplasticity." You are rewiring your own brain. The director isn't just good for your health and important for being effective at work, it's a key ingredient in how you sculpt your brain in the long term.

Putting all this together, all you need to do to change a culture, whether at home or at work, is focus other people's attention in new ways long enough. That's exactly right. But it's also really, really difficult. When Emily asks her kids to change their behavior, they are paying attention, but not to her goal of better communication. They are paying attention to an alarm signal going off in their heads. Sensing someone is trying to change you often creates an automatic threat response, linked to uncertainty, status, and autonomy. As Sir Winston Churchill once said, "I love to learn, but I hate to be taught." If being changed by others is usually a threat, this leads to the idea that when real change occurs, it is probably because an individual has chosen to change his own brain. Self-directed neuroplasticity, the director monitoring and altering the show, may be the true heart of change.

How do you "facilitate self-directed neuroplasticity" on a large scale? There appears to be three key components to this kind of change. First, you need to create a safe environment so that any threat response is minimized. Second, you need to help others focus their attention in just the right ways to create just the right new connections. Finally, to keep any new circuits alive, you have to get people coming back to pay attention to their new circuits over and again.

SAFETY FIRST

Until people's minds are at ease, focusing their attention on your goals is an uphill battle. An effective way to create a sense of safety in the brain is to offer the brain a reward to counteract the threat. You need to find something the brain wants.

Emily's approach was to promise a holiday, hoping the kids would be interested enough in this to be willing to pay attention to her actual goal of improving communication at home. External rewards are often the first solution people grab for, because tangible concepts are easier to hold on the stage than subtler ideas. However, external rewards such as holidays or money have limited use. You can't just keep offering these to motivate people, because if people expect this reward it tends to become less valuable, and a reward isn't so rewarding unless it gets bigger each time, which isn't sustainable.

While your brain doesn't have feelings (it's also dark and quiet in there), it does have its own goals. As you know from the last two acts, your brain ideally likes to feel a sense of increasing status, certainty, autonomy, relatedness, and fairness. In a paper mentioned earlier in this book, called "The Neuroscience of Goal Pursuit," Matt Lieberman and Elliot Berkman write about how external goals (say, a promotion) are evaluated based on how well they align with the brain's intrinsic goals, such as the need for certainty or a sense of autonomy. They call this process assimilation. But why put in an extra step? Why not save time (and perhaps money) and give the brain exactly what it wants?

Emily wants to entice the kids to pay attention to improving communications at home, and reduce the threat of this change with a reward. Rather than promising a holiday, she could have offered the reward of an increase in *status*. Perhaps a reward that treated them as being older or more competent, such as letting them stay up later or watch certain TV shows. In the workplace, you could increase people's status by publicly recognizing them. The positive reward from positive public recognition can resonate with people for years.

To increase *certainty*, Emily could have described what would happen during her proposed family conference, reducing fear of the unknown. In the workplace, increasing a sense of certainty comes from having a better understanding of the big picture. You could reward someone by giving him or her access to more information. Some innovative firms allow all employees access to full financial data, any time they want it. People feel much more certain about their world when they have information, or easy access to it, which puts their mind more at ease and therefore makes them better able to solve difficult problems.

To increase a sense of *autonomy*, Emily could have given the kids the chance to make more of their own decisions—even small ones—such as what to eat for dinner, or when or where they can do their homework. In the workplace, this could be letting people work more flexibly, or work from home, or reducing the amount of reporting required.

To increase *relatedness*, Emily might offer to increase the amount of time the kids spend connecting with their friends, to arrange a party, or to increase their allowed hours for phone calls. In the workplace, an example of this would be giving people opportunities to network with their peers more, by allowing them to attend more conferences or networking groups.

To increase a sense of *fairness*, Emily could have created "fair trades" with the kids: more family bonding time in return for, say, less pressure on them to keep their rooms tidy. In the workplace, some organizations allow employees to have "community days," where they give their time to a charity of their choice. I wonder if helping those in need feels good because of a sense of decreasing unfairness.

Any of the elements of SCARF could have helped Emily reduce Josh and Michelle's sense of threat and create a sense of reward, which would have made it easier to focus their attention in new ways. However, working with the elements of SCARF isn't just about offering tangible rewards. You can leverage the power of the SCARF model in everyday conversations, by paying attention to the way you phrase an idea. If you have a specific task you want someone to do, you might say, "Would you be willing to do this?" rather than "I want you to do this." This simple change takes into account a sense of autonomy.

At times you might use the whole SCARF model, especially when there is a high possible threat level. Imagine you are starting a conversation with a team of people whom you manage, and you want them to pay attention to something difficult. Taking care of status, you might say, "You're all doing great. I'm not here to attack you but to find ways of our becoming even better than we already are." Taking care of certainty, you might say, "I only want to talk for fifteen minutes, and I am not looking for specific outcomes. I just want to hear your ideas." Taking care of autonomy, you might say, "Is that okay with you, if we focus on this right now?" Taking care of relatedness, you might share something about yourself on a human level. Taking care of fairness, you might be careful to point out that you have had the same conversation with everyone else on your team. As you lay all this out, the alarm bells in people's heads start to quiet down, which increases your chances of focusing people's attention in the direction you want.

Business and organizational leaders could benefit from applying the SCARF model much of the time they communicate with others. (Remember that just speaking to someone of a higher status tends to activate a threat.) Many great leaders understand intuitively that they need to work hard to create a sense of safety in others. In this way, great leaders are often humble leaders, thereby reducing the status threat. Great leaders provide clear expectations and talk a lot about the future, helping to increase certainty. Great leaders let others take charge and make decisions, increasing autonomy. Great leaders often have a strong presence, which comes from working hard to be authentic and real with other people, to create a sense of relatedness. And great leaders keep their promises, taking care to be perceived as fair.

On the other hand, ineffective leaders tend to make people feel even less safe, by being too directive, which attacks status. They are not clear with their goals and expectations, which impacts certainty. They micromanage, impacting autonomy, and don't connect on a human level, so there's little relatedness. And they often don't understand the importance of fairness.

Creating a sense of safety is the first step to transforming a culture, whether the culture involves two people at home or twenty thousand at work. SCARF describes the neuroscience basis of the concept psychological safety, an idea that has taken hold in organizations in recent years. It means people feeling challenged but safe enough to perform at their best, speak up, and contribute. Given that any change tends to bring a sense of threat on its own, changing a culture requires creating *toward* states everywhere you can. People will be paying attention either to you or to their fears. The stage isn't big enough for both at once.

FACILITATING THE CREATION OF THE RIGHT CONNECTIONS

Once you have people's attention, next you need to help them focus it in just the right ways. There is an upside to the fact that attention is easily distracted: it's not that hard to distract people from other thoughts and focus them on something new.

A common strategy people use is to tell a story. A good story creates complex maps in the brain as people hold different characters and events on the stage. Stories have some "point," some specific idea at their core, which the storyteller wants others to understand. The point often involves a surprise connection within the story, a character who learns something unexpected. In this way a story might be thought of as an "insight delivery device," a mechanism for having people change their maps.

While some stories can at times be useful, it's all too easy to choose the wrong story, or tell it in the wrong way, or appear insincere if you've told it too often and the story sounds canned. Also, many people know when someone is trying to change them, and when a story gets rolled out, it might make them feel unsafe all over again, dispelling the good work done up front to get their attention in the first place. I know that when someone starts telling me a story, I often have an internal dialogue that says, "Just get to the point," or "Stop trying to convince me of something."

An effective and more direct way to focus attention is simply to ask people the right question, to give them a gap to close. The brain is quite happy closing any gap, as long as it doesn't take too much effort.

Imagine you are a store manager, and you want to change the culture of your team so that people are more focused on customers' needs. Your goal is to ask your team questions that require them to make just the right new connections. The insights from the last scene about facilitating change in individuals apply here, too: the questions should be about solutions, not problems. In a group setting, it's even easier to end up putting too much attention on problems and not enough on solutions.

Going back to the retail store, useful questions a manager could ask his team might include:

> What is one thing you have done that has made a customer delighted in the past?
> What did you do differently that made the customer so happy?
> What would it take for you to do this more often?

These three simple questions could change the behavior of a group more than any long discussion about the challenges of customer service. The questions don't imply a specific answer. They help people arrive at their own insights. These insights could be increased if people were able to discuss the questions in small groups, which reduces status threats and increases a sense of relatedness. When you ask people to answer questions like this, there is an implicit respect inherent in the question, the suggestion that you know people have good answers. It's a status reward, rather than the "what's wrong with us?" question that might threaten status. Most important, solution-focused questions that focus on the exact change you want—improved customer service, in this case—get people to make new connections to increase customer service, rather than focusing on the millions of other details they could focus on. Similar ideas have been fleshed out further in fields such as solutions-focused therapy and appreciative inquiry. I am not proposing that these are brand-new insights. However, I find that it is helpful to have the theoretical explanation for *why* we need to focus attention this way.

In summary, once the general threat level is reduced in a group, focus people's attention on exactly the direction you want them to go in. Remember that the brain is chaotic and easily distracted, so be as clear and specific as possible.

A third way to facilitate wide-scale self-directed neuroplasticity involves establishing goals. When you set a goal, you set up the possibility of a positive (or negative) spiral. Looking out for your goal, you are more likely to perceive information relating to it, which makes you feel positive, because you feel that the goal is going to happen, which makes you look out for it more, and perceive more information, and so on. If the goal involves a positive reward, too, the expectation of a reward can have a strong impact on people's neurochemistry. In this way, if you want people to focus on a change, you might find a way to keep the expectation of a primary reward in sight for as long as possible, as this will lift their moods and improve their thinking.

Setting the right goal can also increase status, by giving people small achievements to notice. The right goal can increase a sense of certainty by providing more clarity on objectives, and it can increase a sense of autonomy if people have a say in how they achieve the goal. Setting the right goal is like a gift that keeps on giving: you continue getting positive benefits all the while you head toward it.

While this is good in theory, unfortunately the goals that people naturally tend to set don't achieve this wave of positive momentum. Jim Barrell, a performance improvement expert working with players from the San Francisco 49ers and the Atlanta Braves, studies the way top performers set goals. "There are toward goals and away goals," Barrell explains, "and which one you use has quite an impact on performance. Toward goals have you visualize and create connections around where you are going. You are creating new connections. What's interesting is you start to feel good at lower levels with toward goals. There are benefits earlier. Away goals have you visualize what can go wrong, which reactivates the emotions involved." The trouble is, because problems come to mind so much easier than solutions, people are always setting away goals instead of toward goals. Also, problems are more certain than unknown solutions, and the brain naturally steers toward certainty. For these reasons and more, toward goals are rare, and setting them might require getting some help from someone else, such as a mentor or coach. The goal Emily tried to set with her family was an away goal: "not to fight." When you set an away goal, you can end up paying attention to the negative emotions instead of making new connections. Lose weight, stop smoking, don't drink: most of the New Year's resolutions of the world are away goals.

There is an additional challenge here to setting goals: the incredible variation across every human being. While brain *processes* are similar (threats diminishing prefrontal cortex resources, for example), what is perceived as a threat, the *content* of thought, has a strong individual component. Thus, when you set goals for other people, not only do you reduce the sense of autonomy, but it's all too easy to think that others are just like you. (To think otherwise takes a lot of space on the stage, and generates uncertainty.) The lesson here: if you are planning on setting goals for other people, perhaps instead create a framework for them to set goals for themselves.

KEEPING NEW CIRCUITRY ALIVE

Once you have reduced threat, and facilitated the right new connections, the third part of the process for changing a culture involves ensuring that people come back to pay attention to their new circuits regularly. If you want a specific new map to stay in place, it's important to reactivate that map regularly. Attention changes the brain, but the brain pays attention to a lot of things. Real change requires repetition.

The term "attention density," coined by Jeffrey M. Schwartz, provides a scientific framework for future research about repeated attention. This density can be measured with such variables as frequency, duration, and intensity of attention. This looks like measuring how often you recall an idea, how long each time, and with how much focus. When you make a promise to another person to do something, it comes into your mind more often and at greater frequency, because of a status threat if you don't do it. The result is the circuits relating to your promise get more attention density, so you are more likely to remember the task. If you write a task down, you are paying far more attention to it than speaking about the task, so again you have increased the attention density, this time through increasing the intensity of the attention itself.

This whole issue is still tricky to study in the lab, because attention is a hard construct to measure. However, there is some good research emerging from the area of learning music, where the significance of repetition comes to light, and from studies on the effect of "rehearsal" on memory encoding, also pointing to the importance of repetition. I have a metaphor for thinking about attention density. Think of the brain as a garden, where it's sunny all the time and rains naturally once in a while. If you want to grow some nice tomatoes, you first plant seedlings, which need careful daily watering. Once the plants are a bit hardy, to keep them growing, you should water them regularly. How often is the right amount? If you water once a year, it will probably wash everything away. Once a quarter won't do much. Once a month will help, maybe. Once a week does make a difference to some plants, but watering twice a week seems to make a sustainable and noticeable difference. It seems the best technique for growing plants is what they do in hydroponic farms, which is to water them with small amounts several times each day. I propose that creating healthy new circuits in the brain is not dissimilar. You need to pay regular attention but it doesn't need to be a lot in one go, it is better if it is regular small amounts.

How do you get other people to pay regular attention to something that's important to you? One of the best ways involves getting them to collaborate. Remember that the brain is eminently social, so if you can get a change you desire linked to the social world, you're on the right track. Creating systems and processes that require people to talk about a project regularly can be as simple as bringing an idea up once a week and having people share their thoughts. Ideas, and brain circuits, come alive in conversation. There are additional benefits to harnessing the power of social interactions. There is a memory network that is activated when information is social that turns out to be more robust than a memory without a social element.

All this points to the need not just to have a strong director yourself, but also to become better at noticing where other people's attention goes. To change a culture, start by paying attention to everyone's attention, and work out how to focus their attention in new ways. Or, better yet, work out ways that other people can activate their own director, to focus their own attention in new ways, and thus rewire their own brains. Learning how to change a culture means learning how to facilitate self-directed neuroplasticity. The more that people can refocus their own attention, the more they can work in synchrony, resonating with the same idea at the same time, just like an orchestra, or a single brain. Perhaps this is what happens when we create change in the world.

LEADING CHANGE

Change is hard, and we urgently need to get better at creating positive change in the world. Unfortunately, many of the people who make it to leadership positions have a highly developed intellect but are poor on the social side of things. Neuroscience is beginning to explore this phenomenon, too. "The brain network involved in working memory, general cognitive problem-solving, goal setting, and planning tends to be on the lateral, or outer, portions of the brain," Matthew Lieberman explains during an interview at his lab. "Then there are regions more involved in the midline or middle areas, related to self-awareness, social cognition, and empathy. We know that these two networks are inversely correlated: when one is active, the other tends to be deactivated. It does suggest possibly that there is something inversely correlated about social and nonsocial abilities." This makes sense when you understand that the networks you pay attention to are the ones that grow. If you spend a lot of time in cognitive tasks, your ability to have empathy with people reduces simply because that circuitry doesn't get used much. Being heavily goal focused, and constantly working conceptually, as most leaders do, will make this situation even worse. There are new studies showing that putting people in positions of power, even a small amount of power, such as being in charge of some team decisions, changes how the brain processes information. In particular, giving people power tends to make them more focused on people as concepts, as tools to use, and less as actual humans with their own intentions and emotions. When we have power, we think less of people as people. This has an upside in your ability to create changes without obsessing about people's feelings but can also blind leaders to the impact of their actions. A leader's impact on others is often quite different from their intent. Put these two factors together and add in the challenges of a high cognitive load, making it hard for people to think well about complex issues, and you have an explanation for why many leaders are strong at driving results, but their people skills often lag. Poor people skills means a poor understanding of your own drivers as well as that of others.

There can be a cost to poor self-knowledge, as Lieberman explains: "There is a study showing that if you show people sentences and you say, 'If we showed you this sentence in a half hour without the last word, would you be able to remember what the word is then?' The extent to which medial prefrontal cortex is active predicts whether or not what they say now is right about what happens later." In this way, leaders who are overly intelligent may make mistakes about their own capacities. And given that the circuitry for knowing yourself is so similar to that for knowing others, there is also a good chance that they will make mistakes about others. Leaders who want to drive change more effectively may want to practice becoming more intelligent about their inner world as a first step. A great way to do this is to discover more about your own brain.

It's time to put all these ideas together and explore how the evening might have gone different for Emily and Paul if they had understood the real drivers of change.

THE CULTURE THAT NEEDS TO TRANSFORM, TAKE TWO

Emily walks through the front door at home, her briefcase overflowing with work to do after dinner. A part of her is looking forward to connecting with the kids, and she notices her disappointment as she sees them ignore her arrival, locked in their own worlds. It could be easy to get upset, but she knows the kids will react badly if she does. She realizes that getting upset but suppressing her emotions won't work, either; the kids will still sense a threat. Emily decides she does want to have a family conversation about how they are getting on, but not to mention it until dinner, when a burst of extra glucose might improve her chances.

It's been a tough day. Emily needs a "little something" to lift her dopamine level until dinner. She decides against a glass of wine, which would only reduce her ability to manage her emotions at dinner, and she phones her mother instead. Her mother is delighted by the unexpected call, and Emily gets to mirror some of her enthusiasm. After half an hour of chatting about little other than the weather and the kids, Emily feels much better.

Paul calls out that dinner is on the table, and the family converges from different parts of the house. Ten minutes after everyone has started eating, Emily launches in with her plan.

"I'd like to have a family conference tonight. Would that be okay with you all?" she asks.

"Oh no, not again, Mom, please. We had one last year," complains Josh.

"Mom, there's nothing to talk about. Everything's fine," Michelle says, her headphones still in one ear.

"Well, let me tell you what I want to discuss, and then you guys can tell me if you are willing to talk." Emily wants the kids to feel more certain and to feel that they have a choice.

Emily planned to launch the discussion with the offer of a reward, thinking that would be enough to make the kids open up, but as she is about to say these words, her director kicks in and senses that this strategy might not work. She needs to get the kids involved in this conversation. She needs to get them to make connections and not just fend off their mother's ideas.

"I want to talk about how we communicate as a family, but I want to do this in a different way, by hearing what you would like to be different."

"I'm in," says Josh.

"And . . . ?" Michelle says, a little more cynical as the older teen.

"Would you be willing to tell me how you want things to be different here?" Emily asks.

"Well . . . ," Michelle pauses. "It's really not cool that you treat me and Josh the same, because I'm much older and more mature than him. I deserve to be treated differently."

Fairness can be big issue at home. Emily was hoping the conversation would take a different tack, focused on her own agenda of connecting more. She has to pause for a moment and consciously let go of her expectations and allow things to take their course. She labels the uncertainty she now feels and decides to be okay with whatever happens.

Paul jumps into the gap that Emily leaves. "What about you, Josh? How do you want things to be different?"

"I want to go to the mall on my own. All my friends do now." Lately Josh has felt his status with his friends slipping, a tough feeling for a young teen male. His parents didn't know this was a problem.

Paul and Emily agree to the kids' requests, with a few provisions, and then Emily asks for a fair exchange in return. "If we do all this, will you both switch off what you are doing when I first get home, just for ten minutes? I used to love seeing you at the end of the day, running to the front door. It helped me feel better after a hard day at work. I'm not asking you to actually be excited to see me, but could we just connect, for ten minutes? We can always make it a snack time together."

"Okay. I'm in," says Josh. The link to another primary reward, food, has gotten his attention.

"And, Michelle, you can tell me what's going on with your friends if you want to. I haven't been patient with you about all that lately. I'm sorry."

Michelle is pleased to know she will be able to talk to her mom about this, even though Emily is overwhelmed by the daily torrent of social updates.

The kids are in a positive state, expecting rewards that are deeply important to each of them. It's a great time to ask a hard question. Emily asks them if they would be willing to work at being kinder to each other, apologizing more when needed, giving each other more help. Ultimately she doesn't want just ten minutes of connecting time. She wants to change the feel of being with her family, the culture. The kids agree that they have let some things slip, and promise to be nicer to each other and to their parents. Small steps are the best way forward. Emily feels that this time, the third attempt to have this conversation, she is going to see some changes.

Over dessert, Emily remembers that she'll need to remind everyone to make sure this new plan works. She gets out a pen and paper and writes up the plan so it's clearer for everyone: ten minutes with Mom at the end of her workday, and being kinder to each other. Paul pipes in that he wants to join in for their "daily ten," as he calls it, if he's home then.

Emily asks the kids how they want to be reminded. Josh wants some stickers made up to put where he wants, and Paul volunteers to help create them on his computer. Michelle wants a note on her cell phone screen that comes up when she turns it on. While Michelle thinks she's being sneaky because her phone is never off, she doesn't know that every time she uses her phone, this reminder will get activated in her brain anyway.

Michelle and Josh finish their last mouthful around the same moment and are about to leap to their rooms, but instead they pause to ask if they can help clean up. In their *toward* state, Michelle and Josh connect more easily to instincts, such as the need for fairness. Emily smiles. They agree to clean up together and then watch a movie as a family. In her clear state of mind, Emily easily predicts that she will do a better job at the work she had planned tonight if she tackles it in the morning, with a fresh mind.

As they watch a funny movie together, the dopamine rush from the humor, combined with surges of oxytocin from shared moments of laughter, combine to relax everyone and make for a great experience. It's been a great day, and despite their differences, the family has bonded and feels as one.

Two hours later, Emily and Paul switch off the television and help their sleepy children get into bed. They whisper about how cute they are when they're asleep. This increases Emily and Paul's attention to their feeling of love for their kids. Feeling warm and connected from the experience tonight, Emily and Paul look downstairs and think for a moment about cleaning up the house more, but almost in unison they make a different choice. They switch the rest of the lights off and walk to their bedroom, closing the door quietly behind them. What goes on in their brain during what happens next . . . well, that's a whole other story.

Surprises About the Brain

- While human change appears hard, change in the brain is constant.
- Focused attention changes the brain.
- Attention goes all too easily to the threat.
- Once you focus attention away from threat, you can create new connections with the right questions.
- Creating long-term change requires paying regular attention to deepen new circuits, especially when they are new.

Some Things to Try

- Practice watching for people's emotional state when you want to facilitate change.
- Don't try to influence people when they are in a strong away state.
- Use elements of the SCARF model to shift people into a toward state.
- Practice using solution-focused questions that focus people's attention directly on the specific circuits you want to bring to life.
- Invent ways to have people pay repeated attention to new circuits.

Encore

The Emily and Paul whom you see at the end of each scene (let's call them Emily and Paul Two) are significantly more effective at their jobs than their counterparts at the start of each scene. But Emily and Paul Two are not just better at managing their email or running meetings. They are also less stressed, have more fun, have a better relationship with their kids, and even appear to have a better sex life. People like this tend be healthier, contribute more to their communities, and even have longer lives.

There was one big difference between the two sets of characters: Emily and Paul Two knew more about their own brains than their counterparts. They had a richer language for the subtle internal signals going on beneath the surface of their attention. This richer language gave them more choices each moment about which mental pathway to follow. Emily and Paul Two had this language because they had strong directors, and having this language also created their strong directors. Their directors could stand apart from and observe their own mental processes. More important, their directors could also make small adjustments, on the fly, to the flow of information in their brains.

The changes in brain functioning that Emily and Paul Two's directors made were tiny, hardly noticeable on brain scans with today's technology. However, that's one of the big insights from this book: microscopic changes in brain functioning, made in a hundredth of a second, can sometimes create massive change in people's lives. This change begins with a slight shift in how energy flows within the brain, perhaps reducing the activation of one brain region and increasing another, and quickly grows into a totally different behavioral response to the same stimuli.

For thousands of years philosophers have said that to "know yourself" was the key to a healthy and successful life. Perhaps what is emerging from the new research about the brain is a new way of thinking about "self-awareness." Only in this case, the "self" is the functioning of your own brain. One of the first things to discover upon exploring the brain is just how much it appears to be like a machine. So much of your mental activity is automatic, driven by forces out of your control, often in reaction to predefined goals, such as maintaining status or certainty. The realization that we are so automatically driven can be frightening to some, but if that is where the story ended, you would be missing a key aspect of being human. While your brain is a machine, it's also not *just* a machine. However, the only way to be more than just a machine is to deeply understand the machine-like nature of your brain. When you begin to know the machine-like nature of your brain, you are building your director. This enables you to say, "That's just my brain," in more situations, which gives you more choices of behaviors. Your capacity to change yourself, change others, and even change the world, may boil down to how well you know your brain, and your capacity to consciously intervene in otherwise automatic processes.

To help distinguish the places where you may now have more choices, let's summarize the insights about the brain that have emerged here. In act 1, you discovered that being able to plan, organize, prioritize, create, or do just about anything except repetitive mental tasks requires using a small, fragile, and energy-hungry brain region, the prefrontal cortex. You discovered the underpinning biology that explains why it's so hard to be in a zone of peak performance and how easily distracted the brain is. You also learned that sometimes the prefrontal cortex is the problem, and you need the ability to shut it down if you want to be more creative. Act 1 was all about learning to work around the limits of your conscious mental processes.

In the intermission you learned about your director, and the importance of being able to step outside your experience and observe your mental functions, which comes from an ability to focus attention in the moment, openly. It became clear that the ability to notice your own mental process in this way has a dramatic impact on your capacity to stop and separate yourself from an automatic train of thought. In other words, you discovered that being able to notice your own thinking process itself was central to knowing and changing your brain.

In act 2, you explored how the brain is built to minimize danger and maximize reward. This occurs as a *toward* and *away* system of emotions, driven by the limbic region of the brain. You saw how *toward* states tend to be more productive for doing good work, but also discovered how easily, quickly, and intensely the *away* state can be. You saw how your thinking capacity can be reduced by remembering past threatening situations, by uncertainty, and by a feeling of lack of autonomy. You discovered two techniques that can help wrestle back control from an overly aroused limbic system: labeling and reappraisal. You also learned about the dramatic impact of expectations on experience. In other words, in act 2 you discovered that your brain's drive to keep you alive sometimes comes with unintended consequences. These consequences can include reducing your mental performance, and can even decreasing your lifespan.

In act 3, you got to see the social world from the brain's perspective, discovering that social domains such as relatedness, fairness, and status can generate either a *toward* or *away* response with the same intensity and using the same circuitry as a reward or danger for one's life. You got to see that a huge amount of human behavior is driven, largely unconsciously, by the desire to minimize social dangers and maximize social rewards.

In act 4, you found out why it's so hard to change other people, because of our natural tendency to focus on problems and make suggestions. You explored a new way of interacting, based on facilitating insights about solutions in others. You looked at what is involved in changing a culture, and explored how the real driver of change is people changing their own brain. You discovered how to help create cultural change by creating a greater sense of safety in ways that deeply impact the brain, then by enabling new connections to occur, and then by helping new circuits to be embedded.

Throughout this book a consistent theme has been the importance of the director. Having a strong director gives you the ability to notice what is happening each moment, instead of acting unconsciously. With a good director you have the capacity to make choices, and these choices change your brain in terms of the neural, mental, and physical behaviors that follow. Over time, your choices can change your brain in much deeper ways, too. It is hoped that through reading this book you have found some innovative ways to build up your own director that suit your lifestyle. Remember that exercises for building your director can be as simple as a few moments of focused attention just before a meal. Repetition is key.

As your *director* becomes stronger, it gets easier to decide what to put on your stage and what to keep off; when to pay close attention to something, and when to step back and allow loose connections to occur instead; how to get decisions onto your stage in the right order, and get them off the stage quickly; how to quiet your mind, to listen to the more subtle signals coming from the two million environmental cues that your brain might be tapping into at any moment, instead of just the forty you can perceive consciously. All of this awaits you in everyday experience. It is hoped that through this book you have enough insight into how your brain functions to give your director things to focus on for years to come.

It may be that understanding the brain is one of the best ways of improving performance in any setting, especially for teams of people working together. As you begin to recognize the patterns outlined in this book, I encourage you to talk about these ideas with others and share any insights you have. The more attention you pay to these concepts, the more of your brain they will take up, and thus the easier it will be to remember these ideas when you need them most. If the ideas in this book exist not just in your head but also in the brains of the people around you, the ideas will be easier to find when you need them. With a good understanding of your brain readily accessible, it will be easier to live a life that is more along the lines of Emily and Paul Two: challenged, but able to use your brain to master those challenges; stretched, but able to grow as a result and achieve great things, whether you are bringing up valuable new members of society, building an innovating new business, or just surviving a tough day at the office.

A final word, a brain-based farewell greeting: May your cortisol levels stay low, your dopamine levels high, your oxytocin run thick and rich, your serotonin build to a lovely plateau, and your ability to watch your brain at work keep you fascinated until your last breath. I wish you well on your journey.

David Rock
January 2008
Somewhere over the Pacific Ocean
between Sydney and Los Angeles

FURTHER RESOURCES

Based on what I have heard over the years, I suspect this book may open a door for many people to an exciting new way of thinking. If that's the case for you, I encourage you to dive in and discover more, and find ways to keep focusing your attention on these insights.

I blog regularly in a number of places, including a website called *Your Brain at Work* at www.your-brain-at-work.com. You will find lots of other scientists and authors from the organization I have developed also writing there. I also blog on *Psychology Today* at www.psychologytoday.com/us/experts/david-rock and regularly for *The Harvard Business Review*, *Quartz*, and others.

You might also like to check out the NeuroLeadership field. There is an annual NeuroLeadership Summit, and a *NeuroLeadership Journal*, which contains articles about the brain in relation to the workplace, focused on leading and managing others. There are links to other educational programs available there, including an online certificate program, if you want to study these ideas more formally. See www.NeuroLeadership.com.

I wrote two earlier books that may be of interest. My previous book, co-authored with Dr. Linda Page, is called *Coaching with the Brain in Mind* (Wiley, 2009). This is a textbook about the brain and related fields such as learning theory and systems theory, for those who want a deeper understanding of theories involved in creating change. My book before that, *Quiet Leadership* (Collins, 2006) explores the science and the art of using conversation to bring out insight in others. This is a great book for people who want to use brain insights to become better leaders, managers, mentors, coaches, teachers, or parents.

If you are interested in building your own leadership or coaching skills further, you could check out the brain-based training programs, which are available around the world. See www.NeuroLeadership .com for more.

To find out more about the school I helped develop, go to www .theblueschool.org.

For information on my work overall, see my blog at www .DavidRock.net.

ACKNOWLEDGMENTS

My biggest thanks go to my wife, Lisa Rock, who, for too long, put up with a husband who traveled a lot and only ever wanted to talk about the brain when he was around. A big thanks to my daughters, India and Trinity, who unfortunately had to practice lots of emotional regulation to let Dad lock himself away and write.

Thanks go to Jeffrey Schwartz, who was originally going to partner with me on this book but decided to go in a new direction midway. Your guidance and mentoring was much appreciated. The terms *self-directed neuroplasticity* and *attention density* came from Jeff. Also thanks to Matt Lieberman, Kevin Ochsner, Evian Gordon, and Yi-Yuan Tang for your informal mentoring and scientific guidance over several years.

Thanks to Al Ringleb, director of CIMBA, the business school in Italy, who helped make all this possible by collaborating on the first *NeuroLeadership Journal* and Summits. Also thanks to Art Kleiner, former editor in chief of the magazine *strategy + business*, for your continuing mentoring and belief. A big thanks to Karen-Jane Eyre, who helped with the editing, and Rachel Sheppard, who helped with organizing the book's reference section. Also thanks to everyone at Harper Business for your support, including CEO Brian Murray, for seeing a spark in me back in 2005.

Big thanks to the thousands of neuroscientists who patiently explore the structure and function of the brain, without whom nothing in this book would have been possible. And a final big, warm, hearty thanks to my brain's own director, without which I wouldn't have completed even the first page of this book.

NOTES

The main ideas in this book are drawn from a handful, up to hundreds, of studies. I have listed the key studies that seem most relevant to each argument, not every paper used to write the book. Some of the papers are available online for free, though a good number also require purchasing to view.

Scene 1: The Morning Information Overwhelm

For more on Roy Baumeister's research on the prefrontal cortex's energy limits, see:

Masicampo, E. J., and R. F. Baumeister. "Toward a physiology of dual-process reasoning and judgment: Lemonade, willpower, and effortful rule-based analysis." *Psychological Science* 19 (2008): 255–60.

Vohs, K. D., R. F. Baumeister, B. J. Schmeichel, J. M. Twenge, N. M. Nelson, and D. M. Tice. "Making choices impairs subsequent self-control: A limited resource account of decision-making, self-regulation, and active initiative." *Journal of Personality and Social Psychology* 94 (2008): 883–98.

For more on different types of memory, see the Atkinson-Shiffrin model, which was proposed in 1968 by Richard Atkinson and Richard Shiffrin. See:

Atkinson, R. C., and R. M. Shiffrin. "Human memory: A proposed system and its control processes." In K. W. Spence and J. T. Spence, eds. *The psychology of learning and motivation* Vol. 2, New York: Academic Press, 1968, pp. 89–195.

Regarding different levels of effort for getting information onto the stage, this idea came from looking at cognitive bias research from psychology and linking it to the effort involved in making decisions. For example, in 1973, psychologists Amos Tversky and Daniel Kahneman explored the idea of the "available heuristic," which is how people think about things that are easy to think about most readily, which tends to be recent thoughts. See:

Tversky, A., and D. Kahneman. "Availability: A heuristic for judging frequency and probability." *Cognitive Psychology* 5 (1973): 207–32.

Our poor abilities at affective forecasting are linked to the difficulty of thinking about subtle factors. We incorrectly estimate what will make us happy in the future because it takes so much effort and energy to picture the future.

See Daniel Gilbert's book *Stumbling on Happiness*, New York: HarperCollins, 2006.

For more on how the brain is wired to think in terms of people interacting, see:

Geary, David C. *The Origin of Mind: Evolution of Brain, Cognition, and General Intelligence.* Washington, D.C.: American Psychological Association, 2004.

Scene 2: A Project That Hurts to Think About

For more on the size of working memory, see:

Miller, G. A. "The magical number seven, plus or minus two: Some limits on our capacity for processing information." *Psychological Review* 63 (1956): 81–97.

Research showing that working memory is more like four items at best includes:

Cowan, N. "The magical number 4 in short-term memory: A reconsideration of mental storage capacity." *Behavioral and Brain Sciences* 24 (2001): 87–185.

Gobet, F., and G. Clarkson. "Chunks in expert memory: Evidence for the magical number four . . . or is it two?" *Memory* 12, no. 6 (2004): 732–47.

Shiffrin, R. M., and R. M. Nosofsky. "Seven plus or minus two: A commentary on capacity limitations." *Psychological Review* 101, no. 2 (1994): 357–61.

More on the timing involved in working memory can be found in:

Baddeley, A. D., N. Thomson, and M. Buchanan. "Word length and the structure of short-term memory." *Journal of Verbal Learning and Verbal Behavior* 14 (1975): 575–89.

Schweickert, R., and B. Boruff. "Short-term memory capacity: Magic number or magic spell?" *Journal of Experimental Psychology: Learning, Memory, and Cognition* 12 (1986): 419–25.

For more background on the soundproof room study, see:

McElree, B. "Working memory and focal attention." *Journal of Experimental Psychology: Learning, Memory, and Cognition* 27, no. 3 (2001): 817–35.

More on cognitive complexity and decision-making can be found in a field called relational complexity. See:

Halford, G., N. Cowan, and G. Andrews. "Separating cognitive capacity from knowledge: A new hypothesis." *Trends in Cognitive Sciences* 11, no. 6 (2007): 236–42.

Halford, G. S., R. Baker, J. McCredden, and J. D. Bain. "How many variables can humans process?" *Psychological Science* 16, no. 1 (2005): 70–76.

For more on Desimone's research on neural competition, see:

Desimone, R. "Visual attention mediated by biased competition in extrastriate visual cortex." *Philosophical Transactions of the Royal Society of London (Biological Sciences)* 353 (1998): 1245–55.

Desimone, R., and J. Duncan. "Neural mechanisms of selective visual attention." *Annual Review of Neuroscience* 18 (1995): 193–222.

Scene 3: Juggling Five Things at Once

For more on Robert Desimone's work on attention, see:

Desimone, R. "Visual attention mediated by biased competition in extrastriate visual cortex." *Philosophical Transactions of the Royal Society of London (Biological Sciences)* 353, (1998): 1245–55.

Desimone, R., and J. Duncan. "Neural mechanisms of selective visual attention." *Annual Review of Neuroscience* 18 (1995): 193–222.

Harold Pashler has numerous papers covering his work on multitasking, bottlenecks, and queues. Some examples of his work include:

Ferreira, V. S., and H. Pashler. "Central Bottleneck Influences on the Processing Stages of Word Production." *Journal of Experimental Psychology: Learning, Memory, and Cognition* 28, no. 6 (2002): 1187–99.

Pashler, H. "Attentional limitations in doing two tasks at the same time." *Current Directions in Psychological Science* 1 (1992): 44–50.

Pashler, H., J. C. Johnston, and E. Ruthruff. "Attention and performance." *Annual Review of Psychology* 52 (2001): 629–51.

For information regarding how health, stress, and status are intertwined, see this information about allostatic load:

Allostatic Load Working Group: Research Network on Socioeconomic Status and Health (1999). Allostatic Load and Allostasis. Retrieved from http://www.macses.ucsf.edu/Research/Allostatic/notebook/allostatic.html (accessed on April 10, 2009).

The University of London study about multitasking and reduced IQ was reported by Dr. Glenn Wilson, a psychologist at King's College, London. The study was sponsored by Hewlett-Packard, and not formally published as a paper. There has been some controversy about the paper, as some media outlets incorrectly shared the data.

For more on the importance of paying close attention to information in order to form long-term memories, see:

Ezzyat, Y, and L. Davachi. "The influence of event perception on long-term memory formation." Delivered at the Fifteenth Annual Meeting of the Cognitive Neuroscience Society, San Francisco, Calif., April 2008.

The basal ganglia is a major brain region. There is even an International Basal Ganglia Society. (See: http://www.ibags.info/.) Ann M. Graybiel is an important researcher in this area, with research in her laboratory focused on regions of the forebrain that influence movement, mood, and motivation: the basal ganglia and neural pathways interconnecting the basal ganglia with the cerebral cortex.

For more on Gerald Edelman's work on neural Darwinism, see his book *Brilliant Air, Brilliant Fire*, New York: Basic Books, 1993.

For more on how repetitive tasks cause long-term potentiation or "embedding" in the brain, see:

Bodner, M., Y. Zhou, G. L. Shaw, and J. M. Fuster. "Symmetric temporal patterns in cortical spike trains during performance of a short-term memory task." *Neurological Research* 19 (1997): 509–14.

For more on the study using the keyboard and noticing patterns unconsciously, see:

Rauch, S. L., C. R. Savage, H. D. Brown, T. Curran, N. M. Alpert, A. Kendrick, A. J. Fischman, and S. M. Kosslyn. "A PET Investigation of Implicit and
Explicit Sequence Learning." *Human Brain Mapping* 3 (1995): 271–86.

Scene 4: Saying No to Distractions

The study on office distractions is from Basex, a New York research firm.
The study's twenty-six-page report, called "Information Overload: We Have
Met the Enemy and He Is Us," looks at strategies companies can use to cope with
information overload, including ten tips designed to ease the burden immediately. The study is available only by purchase from www.basex.com.

The data regarding changes in attention and Microsoft's efforts to reduce
distractions appeared in Clive Thompson, "Meet the Life Hackers," *New York
Times*, October 16, 2005.

For more on how your smartphone influences IQ even if it is off but still in
the room, see:

Meyer, Robinson. "Your Smartphone Reduces Your Brainpower, Even If It's
Just Sitting There." *The Atlantic*, August 2, 2017. Retrieved from https://www
.theatlantic.com/technology/archive/2017/08/a-sitting-phone-gathers-brain
-dross/535476/.

For information regarding ambient neural activity, see:

Hedden, T., and J. D. Gabrieli. "The ebb and flow of attention in the human
brain." *Nature Neuroscience* 9 (2006): 863–65.

For more on how schizophrenia may reduce people's ability to inhibit task-
irrelevant signals, see Amy Arnsten's work on the prefrontal cortex, including:

Arnsten, A.F.T. "Catecholamine and second messenger influences on prefrontal cortical networks of 'representational knowledge': A rational bridge between
genetics and the symptoms of mental illness." *Cerebral Cortex* 18 (2007): i6–i15.

Vijayraghavan, S., M. Wang, S. G. Birnbaum, G. V. Williams, and A.F.T.
Arnsten. "Inverted-U dopamine D1 receptor actions on prefrontal neurons engaged in working memory." *Nature Neuroscience* 10 (2007): 376–84.

For more on people's ability not to think of a task, see:

Wegner, D. M., D. J. Schneider, S. Carter III, and T. L. White. "Paradoxical
effects of thought suppression." *Journal of Personality and Social Psychology*
53, no. 1 (1987): 5–13.

For more on lapses in attention and on activation of the medial prefrontal
cortex, see:

Mason, M. F., M. I. Norton, J. D. Van Horn, D. M. Wegner, S. T. Graf-
ton, and C. N. Macrae. "Wandering minds: The default network and stimulus-
independent thought." *Science* 315 (2007): 393–95.

The insights for the idea that "bad is stronger than good" come from two
places. First from Jonathan Haidt's *The Happiness Hypothesis*, New York: Basic
Books, 2005; and from the paper referenced here:

Baumeister, R. F., E. Bratslavsky, C. Finkenauer, and K. D. Vohs. "Bad is stronger than good." *Review of General Psychology* 5, no. 4 (2001): 323–70.

For more on how the orbital frontal cortex detects changes in expectations and increased novelty, see:

Leung, H., P. Skudlarski, J. C. Gatenby, B. S. Peterson, and J. C. Gore. "An event-related functional MRI study of the Stroop color word interference task." *Cerebral Cortex* 10, no. 6 (2000): 552–60.

MacLeod, C. "Half a century of research on the Stroop effect: An integrative review." *Psychological Bulletin* 109 (1991): 163–203.

Petrides, M. "The orbitofrontal cortex: Novelty, deviation from expectation, and memory." *Annals of the New York Academy of Sciences* 1121 (2007): 33–53.

For more on the right ventrolateral prefrontal cortex, see:

Lieberman, M. D., N. I. Eisenberger, M. J. Crockett, S. M. Tom, J. H. Pfeifer, & B. M. Way. "Putting feelings into words: Affect labeling disrupts amygdala activity in response to affective stimuli." *Psychological Science* 18, no. 5 (2007): 421–28.

For more about dopamine and arousal, see:

Schultz, W. "The reward signal of midbrain dopamine neurons." *News in Physiological Sciences* 14, no. 6 (1999): 249–55.

———. "Reward signaling by dopamine neurons." *Neuroscientist* 7, no. 4 (2001): 293–302.

Waelti, P., A. Dickinson, and W. Schultz. "Dopamine responses comply with basic assumptions of formal learning theory." *Nature* 412 (2001): 43–48.

To view Baumeister's self-control study, see:

Gailliot, M. T., R. F. Baumeister, C. N. DeWall, J. K. Maner, E. A. Plant, D. M. Tice, L. E. Brewer, and B. J. Schmeichel. "Self-control relies on glucose as a limited energy source: Willpower is more than a metaphor." *Journal of Personality and Social Psychology* 92, no. 2 (2007): 325–36.

For more on the work of Jonathan Haidt, see:

Haidt, J. *The Happiness Hypothesis: Finding Modern Truth in Ancient Wisdom.* New York: Basic Books, 2005.

For more on Benjamin Libet, see:

Libet, B., E. W. Wright, B. Feinstein, and D. Pearl. "Subjective referral of the timing for a conscious sensory experience: A functional role for the somatosensory specific projection system in man." *Brain* 102, no. 1 (1979): 193–224.

The idea of having "free won't" was introduced by Jeffrey Schwartz in his book *The Mind and the Brain*, New York: Harper Perennial, 2003.

For more on explicit versus implicit awareness, see Matthew Lieberman's work on intuition:

Lieberman, M. D. "Intuition: A social cognitive neuroscience approach." *Psychological Bulletin* 126 (2000): 109–37.

Also see Rauch's study mentioned in scene 3:

Rauch, S. L., C. R. Savage, H. D. Brown, T. Curran, N. M. Alpert, A. Kendrick, A. J. Fischman, and S. M. Kosslyn. "A PET investigation of implicit and explicit sequence learning." *Human Brain Mapping* 3 (1995): 271–86.

Scene 5: Searching for the Zone of Peak Performance

The Yerkes-Dodson Law defines the relationship between arousal and performance. It was originally observed by Robert M. Yerkes and John Dillingham Dodson in a paper published in 1908.

Yerkes, R. M., and J. D. Dodson. "The relation of strength of stimulus to rapidity of habit-formation." *Journal of Comparative Neurology and Psychology* 18 (1908): 459–82

For more on how stress impacts performance, see:

Arnsten, A.F.T. "The biology of being frazzled." *Science* 280 (1998): 1711–12.

Mather, M., K. J. Mitchell, C. L. Raye, D. L. Novak, E. J. Greene, and M. K. Johnson. "Emotional arousal can impair feature binding in working memory." *Journal of Cognitive Neuroscience* 18 (2006): 614–25.

Vijayraghavan, S., M. Wang, S. G. Birnbaum, G. V. Williams, and A.F.T. Arnsten. "Inverted-U dopamine D1 receptor actions on prefrontal neurons engaged in working memory." *Nature Neuroscience* 10 (2007): 376–84.

For more on dopamine and noradrenalin levels and good prefrontal cortex functioning, see:

Birnbaum, S. G., P. X. Yuan, M. Wang, S. Vijayraghavan, A. K. Bloom, D. J. Davis, K. T. Gobeske, J. D. Sweatt, H. K. Manji, and A.F.T. Arnsten (2004). "Protein kinase C overactivity impairs prefrontal cortical regulation of working memory." *Science* 306, no. 5697 (2004): 882–84.

Vijayraghavan, S., M. Wang, S. G. Birnbaum, G. V. Williams, and A.F.T. Arnsten. "Inverted-U dopamine D1 receptor actions on prefrontal neurons engaged in working memory." *Nature Neuroscience* 10 (2007): 376–84.

For more on the link between fear and cognition, see:

Phelps, E. A. "Emotion and cognition: Insights from Studies of the Human Amygdala." *Annual Review of Psychology* 57 (2006): 27–53.

For more on the study about increasing muscle mass through visualization, see:

Yue, G., and K. J. Cole. "Strength increases from the motor program: Comparison of training with maximal voluntary and imagined muscle contracts." *Journal of Neurophysiology* 67 (1992): 1114–23

For more on the impact of visualization processes, see:

Robertson, Ian. *Opening the Mind's Eye: How Images and Language Teach Us How to See.* New York: St. Martin's Press, 2003.

For more on dopamine and love, see:

Aron A., H. Fisher, D. J. Mashek, G. Strong, H. Li, and L. L. Brown. "Reward, motivation, and emotion systems associated with early-stage intense romantic love." *Journal of Neurophysiology* 94 (2005): 327–37.

Fisher, H. *Why We Love: The Nature and Chemistry of Romantic Love.* New York: Henry Holt and Company, 2004.

For more on how arousal is individual, see:

Coghill, R. C., J. G. McHaffie, Y. Yen. "Neural correlates of inter-individual differences in the subjective experience of pain." *Proceedings of the National Academy of Sciences*, 100 (2003): 8538–42.

Shansky, R. M., C. Glavis-Bloom, D. Lerman, P. McRae, C. Benson, K. Miller, L. Cosand, T. L. Horvath, and A.F.T. Arnsten. "Estrogen mediates sex differences in stress-induced prefrontal cortex dysfunction." *Molecular Psychiatry* 9 (2004): 531–38.

For more on the three types of happiness, see *Authentic Happiness*, by Martin Seligman, New York: Free Press, 2005.

Scene 6: Getting Past a Roadblock

For more on priming, see:

Jacoby, L. L. (1983). "Perceptual Enhancement: Persistent Effects of an Experience." *Journal of Experimental Psychology: Learning, Memory, and Cognition* 9, no. 1 (1983): 21–38.

Impasse theory was developed by Stellan Ohlsson. See:

Knoblich, G., S. Ohlsson, H. Haider, and D. Rhenius. (1999). "Constraint relaxation and chunk decomposition in insight problem solving." *Journal of Experimental Psychology: Learning, Memory, and Cognition* 25, no. 6 (1999): 1534–55

For more of Richard Florida's work, see his book:

Florida, R., *The Rise of the Creative Class*. New York: Basic Books, 2002.

For more on novelty, see:

Petrides, M. "The orbitofrontal cortex: Novelty, deviation from expectation, and memory. *Annals of the New York Academy of Sciences* 1121 (2007): 33–53.

Dr. Mark Jung-Beeman has several great papers. For a good summary of his work see:

Bowden, E. M., M. Beeman, J. Fleck, and J. Kounios. "New approaches to demystifying insight." *Trends in Cognitive Sciences* 9 (2005): 322–28.

For more on how anxiety and positive mood impact insight, see:

Subramaniam, K., J. Kounios, E. M. Bowden, T. B. Parrish, and M. Beeman. "Positive mood and anxiety modulate anterior cingulate activity and cognitive preparation for insight." *Journal of Cognitive Neuroscience*, in press.

For more on the brain frequencies required for insight, see:

Kounios, J., J. I. Fleck, D. L. Green, L. Payne, J. L. Stevenson, E. M. Bowden, and M. Beeman. "The origins of insight in resting-state brain activity." *Neuropsychologia* 46 (2008): 281–91.

Kounios, J., J. L. Frymiare, E. M. Bowden, J. I. Fleck, K. Subramaniam, T. B. Parrish, and M. Beeman. "The prepared mind: Neural activity prior to problem presentation predicts solution by sudden insight." *Psychological Science* 17 (2006): 882–90.

For more on the right hemisphere and insight, see:

Bowden, E. M., and M. Beeman. "Aha! Insight experience correlates with solution activation in the right hemisphere." *Psychonomic Bulletin and Review* 10 (2003): 730–37.

Jonathon Schooler's idea of the "a-duh" experience was first published in the *Journal of Experimental Psychology*:

Dougal, S., and J. W. Schooler. "Discovery misattribution: When solving is confused with remembering." *Journal of Experimental Psychology* 136, no. 4 (2007): 577–92

The ARIA model was outlined in my book *Quiet Leadership*, New York: Collins, 2006. It was first appeared in an academic journal here:

Rock. D., "A brain based approach to coaching," *The International Journal of Coaching in Organizations* 4, no. 2 (2006): 32–43.

For more on Jonathan Schooler's work on verbalizing interfering with insight, see:

Schooler, J. W., S. Ohlsson, and K. Brooks. "Thoughts beyond words: When language overshadows insight." *Journal of Experimental Psychology* 122, no. 2 (1993): 166–83.

The information regarding 75 percent of people solving insights is a summary of data collected at dozens of workshops over three years. The highest number was 100 percent, and the lowest around 50 percent. Most of the time a group of people achieved 75 percent or higher.

For more on the effects of mindfulness on well-being and performance, see:

Hassed, C. "Mindfulness, wellbeing, and performance." *NeuroLeadership Journal* 1 (2008): 53–60.

Intermission: Meet the Director

You can explore the idea of an episodic buffer further in this paper:

Baddeley, A. "The episodic buffer in working memory." *Trends in Cognitive Sciences* 4, no. 11 (2000): 417–23.

For more on the way the prefrontal cortex manages the overall brain, see:

Miller, E. K., and J. D. Cohen. "An integrative theory of prefrontal cortex function." *Annual Review of Neuroscience* 24 (2001): 167–202.

For an introduction to social cognitive neuroscience, illustrating the early ideas of the field, see:

Ochsner, K. N., and M. D. Lieberman. "The emergence of social cognitive neuroscience." *American Psychologist* 56 (2001): 717–34.

A summary of definitions of mindfulness can be found here:

Bishop, S. R., M. Lau, S. Shapiro, L. Carlson, N. D. Anderson, J. Carmody, Z. V. Segal, S. Abbey, M. Speca, D. Velting, and G. Devins. "Mindfulness: A proposed operational definition." *Clinical Psychology: Science and Practice* 11, no. 3 (2004): 230–41.

For more on Kirk Brown's Mindful Awareness Attention Scale and how mindfulness allows people to connect to subtler internal signals, see:

Brown, K. W., and R. M. Ryan. "The benefits of being present: Mindfulness and its role in psychological well-being." *Journal of Personality and Social Psychology* 84, no. 4 (2003): 822–48.

For more on Jon Kabat-Zinn's studies of mindfulness helping with skin disease recovery, see:

Kabat-Zinn, J., E. Wheeler, T. Light, A. Skillings, M. J. Scharf, T. G. Cropley, D. Hosmer, and J. D. Bernhard. (1998). "Influence of a mindfulness meditation-based stress reduction intervention on rates of skin clearing in patients with moderate to severe psoriasis undergoing phototherapy (UVB) and photochemotherapy (PUVA)." *Psychosomatic Medicine* 60, no. 5 (1998): 625–32.

For more on mindfulness and immune function, see:

Davidson, R. J., J. Kabat-Zinn, J. Schumacher, M. Rosenkranz, D. Muller, S. F. Santorelli, F. Urbanowski, A. Harrington, K. Bonus, and J. F. Sheridan. "Alterations in brain and immune function produced by mindfulness meditation." *Psychosomatic Medicine* 65, no. 4 (2003): 564–70.

Mark Williams's research, and more on mindfulness and depression, can be found in this book:

Williams, M., J. D. Teasdale, Z. V. Segal, and J. Kabat-Zinn. *The Mindful Way Through Depression: Freeing Yourself from Chronic Unhappiness.* New York: The Guilford Press, 2007.

A good paper on mindfulness and depression is:

Teasdale, J. D., M. Pope, and Z. V. Segal. "Metacognitive Awareness and Prevention of Relapse in Depression: Empirical Evidence." *Journal of Consulting and Clinical Psychology* 70, no. 2 (2002): 275–87.

For more on Yi-Yuan Tang's study comparing mindfulness training to relaxation training, see:

Tang, Y. Y., and M. I. Posner. "The neuroscience of mindfulness." *Neuro-Leadership Journal* 1 (2008): 33–37.

Tang Y. Y., Y. Ma, J. Wang, Y. Fan, S. Feng, Q. Lu, Q. Yu, D. Sui, M. K. Rothbart, M. Fan, and M. I. Posner. "Short-term meditation training improves attention and self-regulation." *Proceedings of the National Academy of Sciences* 104, no. 43 (2007): 17152–56.

Studies of mindfulness and gamma activity include:

Kaiser, Jochen, and W. Lutzenberger. "Human gamma-band activity: A window to cognitive processing." *NeuroReport* 16, no. 3 (2005): 207–11.

Lutz, A., L. L. Greischar, N. B. Rawlings, M. Ricard, and R. J. Davidson. "Long-term meditators self-induce high-amplitude gamma synchrony during mental practice." *Proceedings of the National Academy of Sciences* 101, no. 46 (2004): 16369–73.

For more on mindfulness and cognitive control, see:

Brefczynski-Lewis, J. A., A. Lutz, H. S. Schaefer, D. B. Levinson, and R. J. Davidson. "Neural correlates of attentional expertise in long-term meditation practitioners." *Proceedings of the National Academy of Sciences* 104, no. 27 (2003): 11483–88.

Creswell, J. D., B. M. Way, N. I. Eisenberger, and M. D. Lieberman. (2007). "Neural correlates of dispositional mindfulness during affect labeling." *Psychosomatic Medicine* 69 (2007): 560–65.

Kaiser, Jochen, and W. Lutzenberger. "Human gamma-band activity: A window to cognitive processing." *NeuroReport* 16, no. 3 (2005): 207–11.

Posner, M. I., M. K. Rothbart, B. E. Sheese, and Y. Y. Tang. "The anterior cingulate gyrus and the mechanism of self-regulation." *Cognitive, Affective and Behavioral Neuroscience* 7, no. 4 (2007): 391–95.

The study of mindfulness in couples can be found here:

Barnes, S., K. W. Brown, E. Krusemark, K. W. Campbell, and R. D. Rogge. "The role of mindfulness in romantic relationship satisfaction and responses to relationship stress." *Journal of Marital and Family Therapy* 33, no. 4 (2007): 482–500.

For more on the Farb paper exploring two states of experience, see:

Farb, N.A.S., Z. V. Segal, H. Mayberg, J. Bean, D. McKeon, Z. Fatima, and A. K. Anderson. "Attending to the present: Mindfulness meditation reveals distinct neural modes of self-reference." *Social Cognitive Affective Neuroscience* 2 (2007): 313–22.

A good discussion of the Farb paper by Daniel J. Siegel is here:

Siegel, D. J. "Mindfulness training and neural integration: differentiation of distinct streams of awareness and the cultivation of well-being." *Social Cognitive Affective Neuroscience* 2, no. 4 (2007): 259–63.

http://www.pubmedcentral.nih.gov/articlerender.fcgi?artid=2566758—FN1

For more on the medial prefrontal cortex and knowing yourself, see:

Amodio, D. M., and C. D. Frith. "Meeting of minds: the medial frontal cortex and social cognition." *Nature Reviews Neuroscience* 7 (2004): 268–77.

Gusnard, D.A., E. Akbudak, G. L. Shulman, and M. E. Raichle. "Medial prefrontal cortex and self-referential mental activity: Relation to a default mode of brain function." *Proceeding of the National Academy of Sciences* 98 (2001): 4259–64.

Macrae, C. N., J. M. Moran, T. F. Heatherton, J. F. Banfield, and W. M. Kelley. "Medial prefrontal activity predicts memory for self." *Cerebral Cortex* 14 (2004): 647–54.

For more on interoception, see:

Craig A. D. "How do you feel? Interoception: the sense of the physiological condition of the body." *National Review of Neuroscience* 3 (2002): 655–66.

A good summary of all the research on mindfulness and its impact on health is:

Brown, K. W., and R. M. Ryan. "Mindfulness: Theoretical foundations and evidence for its salutary effects." *Psychological Inquiry* 18, no. 4 (2007): 211–37.

Also see:

Davidson, R. J., J. Kabat-Zinn, J. Schumacher, M. Rosenkranz, D. Muller, S. F. Santorelli, F. Urbanowski, A. Harrington, K. Bonus, and J. F. Sheridan. "Alterations in brain and immune function produced by mindfulness meditation." *Psychosomatic Medicine* 65, no. 4 (2003): 564–70.

For more on John Teasdale's work, see:

Teasdale, J. D. (1999). "Metacognition, mindfulness, and the modification of mood disorders." *Clinical Psychology and Psychotherapy* 6 (1999): 146–55.

For more on Daniel Siegel and mindfulness, see his book:

Siegel, D. J. *The Mindful Brain: Reflection and Attunement in the Cultivation of Well-being.* New York: W. W. Norton and Company, 2007.

The term *attention density* was coined by Dr. Jeffrey M. Schwartz, in this paper:

Schwartz, J. M., H. P. Stapp, and M. Beauregard. "Quantum physics in neuroscience and psychology: A neurophysical model of mind-brain interaction." *Philosophical Transactions of the Royal Society*, 2005. Published online, doi:10.1098/rsub200401598, 2005; http://rstb.royalsocietypublishing.org/content/360/1458/1309.abstract.

Research on how mindfulness can change the brain in the long term includes:

Lazar, S. W., C. E. Kerr, R. H. Wasserman, J. R. Gray, D. N. Greve, M. T. Treadway, M. McGarvey, B. T. Quinn, J. A. Dusek, H. Benson, S. L. Rauch, C. I. Moore, B. Fischl. "Meditation experience is associated with increased cortical thickness." *Neuroreport* 16, no. 17 (2005): 1893–97.

Schwartz, J. M. "A role for volition and attention in the generation of new brain circuitry: Toward a neurobiology of mental force. *Journal of Consciousness Studies* 6, no. 8–9 (1999): 115–42.

Scene 7: Derailed by Drama

For more on the structure of the limbic system, see:

LeDoux, J. *The Emotional Brain: The Mysterious Underpinnings of Emotional Life*. New York: Simon and Schuster, 1998.

For more on Evian Gordon's work on the brain being organized to minimize threat and maximize reward, see Gordon's integrative neuroscience theory:

Gordon, E., ed. *Integrative Neuroscience: Bringing Together Biological, Psychological and Clinical Models of the Human Brain*. Singapore: Harwood Academic Publishers, 2000.

Gordon, E. and L. Williams et al. "An 'integrative neuroscience' platform: applications to profiles of negativity and positivity bias." *Journal of Integrative Neuroscience* 7, no. 3 (2008): 345–66.

For more on the approach/avoid (toward and away) system, there is an entire edited volume of studies. See:

Elliot, A., ed. *Handbook of Approach and Avoidance Motivation*. London: Psychology Press, 2008.

The introductory paper in this volume is helpful for more context on the toward and away system. See:

Elliot, A., "Approach and Avoidance Motivation." *Handbook of Approach and Avoidance Motivation*. London: Psychology Press, 2008.

For more on how we automatically classify stimuli as reward or threat, see:

Fazio, R. H. "On the automatic activation of associated evaluations: An overview." *Cognition and Emotion* 15 (2001): 115–41.

For more on the study of nonsense words activating a threat response, see:

Naccache, L., R. L. Gaillard, C. Adam, D. Hasboun, S. Clemenceau, M. Baulac, S. Dehaene, and L. Cohen. "A direct intracranial record of emotions evoked by subliminal words." *Proceedings of the National Academy of Science* 102 (2005): 7713–17.

For more on the amygdala see:

Phelps, E. A. "Emotion and cognition: Insights from studies of the human amygdala." *Annual Review of Psychology* 57 (2006): 27–53.

For more on bad being stronger than good, see:

Baumeister, R. F., E. Bratslavsky, C. Finkenauer, and K. D. Vohs. "Bad is stronger than good." *Review of General Psychology* 5, no. 4 (2001): 323–70.

Hot spots is a term I coined in 2001 as part of a framework called the Clarity of Distance, which was published in *Quiet Leadership*, New York: Harper-Collins, 2006. I had noticed that people became unable to think clearly in a number of situations, including when an emotional issue arose.

For more on the neuroscience of emotional arousal, see:

Mather, M., K. J. Mitchell, C. L. Raye, D. L. Novak, E. J. Greene, and M. K. Johnson. "Emotional arousal can impair feature binding in working memory." *Journal of Cognitive Neuroscience* 18 (2006): 614–25.

The study of the maze with the owl or cheese can be found in:

Friedman, R. S., and J. Förster. "The effects of promotion and prevention cues on creativity." *Journal of Personality and Social Psychology* 81, no. 6 (2001): 1001–13.

For more on accidental connections, generalizing, and other functions of the limbic system, see:

LeDoux, J. *The Emotional Brain: The Mysterious Underpinnings of Emotional Life.* New York: Simon and Schuster, 1998.

For more on James Gross's model, including the timing of emotions, see:

Ochsner K. N., and J. J. Gross. "The cognitive control of emotion." *Trends in Cognitive Sciences* 9, no. 5 (2005): 242–49.

A study comparing suppression and reappraisal in detail, including the brain regions and timing involved in both, can be found in:

Goldin, P. R., K. McRae, W. Ramel, and J. J. Gross. "The neural bases of emotion regulation: Reappraisal and suppression of negative emotion." *Biological Psychiatry* 63 (2008): 577–86.

For more on Gross's study on suppression impacting memory, see:

Richards, J. M., and J. J. Gross. "Personality and emotional memory: How regulating emotion impairs memory for emotional events." *Journal of Research in Personality* 40, no. 5 (2006): 631–51.

For more on the study comparing suppression to reappraisal outside the lab, see:

Gross, J. J., and O. P. John. "Individual differences in two emotion regulation processes: Implications for affect, relationships, and well-being." *Journal of Personality and Social Psychology* 85, no. 2 (2003): 348–62.

For more on labeling emotions and how this dampens limbic system arousal, see:

Lieberman, M. D., N. I. Eisenberger, M. J. Crockett, S. M. Tom, J. H. Pfeifer, & B. M. Way. "Putting feelings into words: Affect labeling disrupts amygdala activity in response to affective stimuli." *Psychological Science* 18, no. 5 (2007): 421–28.

For more on the study by Lieberman showing that asking people to predict their labeling of emotions can make the emotion worse, see:

Lieberman, M.D., T. Inagaki, M. Crockett, and G. Tabibnia. "Affect labeling is a form of incidental emotion regulation: Subjective experience during affect labeling, reappraisal, and distraction, forthcoming.

For information regarding how health, stress, and status are intertwined, see the following on allostatic load:

Allostatic Load Working Group: Research Network on Socioeconomic Status and Health (1999). Allostatic Load and Allostasis. Retrieved from http://www.macses.ucsf.edu/Research/Allostatic/notebook/allostatic.html (accessed on April 10, 2009).

For more on David Creswell's research exploring the level of activity in the brain's braking system and how this corresponds with people's level of mindfulness, see:

Creswell, J. D., B. M. Way, N. I. Eisenberger, and M. D. Lieberman. "Neural correlates of dispositional mindfulness during affect labeling." *Psychosomatic Medicine* 69 (2007): 560–65.

Scene 8: Drowning amid Uncertainty

For Jeff Hawkins's book about the cortex and prediction, see:

Hawkins, J., and S. Blakeslee. *On Intelligence.* New York: Times Books, 2004.

For more on the impact of uncertainty generating a threat response, see:

Darnon, C., J. M. Harackiewicz, F. Butera, G. Mugny, and A. Quiamzade. "Performance-approach and performance-avoidance goals: When uncertainty makes a difference." *Personality and Social Psychology Bulletin* 33, no. 6 (2007): 813–27.

Hsu, M., M. Bhatt, R. Adolphs, D. Tranel, and C. F. Camerer. "Neural systems responding to degrees of uncertainty in human decision-making." *Science* 310 (2005): 1681–83.

For more on Steve Maier's work, see:

Maier, S. F., R. C. Drugan, and J. W. Grau. "Controllability, coping behavior, and stress-induced analgesia in the rat." *Pain* 12 (1982): 47–56.

Steve Maier also worked with Martin Seligman and others to develop the idea of learned helplessness, which can happen when you feel you have no control over a stressor. For more on learned helplessness, see the following book:

Seligman, M. *Learned Optimism: How to Change Your Mind and Your Life.* Sydney: Random House Publishers, 1992.

For more on Steven Dworkin's study of rats and autonomy, see:

Dworkin, S. I., S. Mirkis, and J. E. Smith. "Response-dependent versus response-independent presentation of cocaine: Differences in the lethal effects of the drug. *Psychopharmacology* 117 (1995): 262–66.

For more on autonomy and control, see:

Mineka, S., and R. W. Hendersen. "Controllability and predictability in acquired motivation." *Annual Review of Psychology* 36 (1985): 495–529.

For more on the study of British civil servants regarding the health and mortality effects of social position in the workplace and the amount of control a person feels they have over his job, see:

Marmot, M., H. Bosma, H. Hemingway, E. Brunner, and S. Stansfeld. "Contribution of job control and other risk factors to social variations in coronary heart disease incidence." *The Lancet* 350 (1997): 235–39.

For an example of a study about people starting small business wanting more "work-life balance," see:

The 2007 MYOB Special Focus Report into the lifestyle of Small Business Owners. This can be downloaded from the MYOB website, under "About MYOB>>News>>MYOB Small Business Surveys>>Survey Special Focus Report-December 2007," at www.myob.com.au.

For more on the study regarding the health and longevity benefits to nursing homes residents of providing simple choices, see:

Rodin, J., and E. J. Langer. "Long-term effects of a control-relevant intervention with the institutionalized aged." *Journal of Personality and Social Psychology* 33, no. 12 (1977): 897–902.

For more on long-term studies and well-being, see:

Diener, E., W. Ng, J. Harter, and R. Arora. "Wealth and happiness across the world: Material prosperity predicts life evaluation, whereas psychosocial prosperity predicts positive feeling." *Journal of Personality and Social Psychology* 99, no. 1 (2010): 52–61.

For a workplace study on autonomy, see: Wood, S., and L.M. de Menezes. "High involvement management, high-performance work systems and well-being." *The International Journal of Human Resource Management* 22, no. 7 (2011): 1586–1610.

For more on the teen brain, and how teenagers in the United States have half the choices of a felon in prison, see:

Epstein, Robert. *The Case Against Adolescence: Rediscovering the Adult in Every Teen*. Fresno, Calif.: Quill Driver Books, 2007.

For more on reappraisal, see:

Goldin, P. R., K. McRae, W. Ramel, and J. J. Gross. "The neural bases of emotion regulation: Reappraisal and suppression of negative emotion." *Biol Psychiatry* 63, no. 6 (2008): 577–86.

Ochsner, K. N., R. D. Ray, J. C. Cooper, E. R. Robertson, S. Chopra, J.D.E. Gabrieli, et al. "For better or for worse: Neural systems supporting the cognitive down and up-regulation of negative emotion." *Neuroimage* 23, no. 2 (2004): 483–99.

For more on how suppressing emotions impacts others, see:

Butler, E. A., B. Egloff, F. H. Wilhelm, N. C. Smith, E. A. Erickson, and J. J. Gross. "The social consequences of expressive suppression." *Emotion* 3, no. 1 (2003): 48–67.

For more on teenagers becoming more capable of cognitive change each year as they get older, see:

Steinberg, L.. "A social neuroscience perspective on adolescent risk-taking." *Developmental Review* 28, no. 1 (2008): 78–106.

Walter Freeman's statement "All the brain can know it knows from within itself" comes from his book *How Brains Make Up Their Minds*, New York: Columbia University Press, 2001.

Scene 9: When Expectations Get Out of Control

For more on expectations altering neuron functioning, see:

Lauwereyns, J., Y. Takikawa, R. Kawagoe, S. Kobayashi, M. Koizumi, B. Coe, M. Sakagami, and O. Hikosaka, "Feature-based anticipation of cues that predict reward in monkey caudate nucleus, *Neuron* 33, no. 3 (January 31, 2002): 463–73.

For more on goals, see:

Berkman, E., and M. D. Lieberman. "The neuroscience of goal pursuit: Bridging gaps between theory and data." In G. Moskowitz and H. Grant, eds. *The Psychology of Goals*. New York: Guilford Press, 2009, pp. 98–126.

Elliot, Andrew, ed. *Handbook of Approach and Avoidance Motivation*. London: Psychology Press, 2008.

For more on how expectations impact experience, see:

Hansen, T., M. Olkonnen, S. Walter, and K. R. Gegenfurtner. "Memory Modulates Color Appearance." *Nature Neuroscience* 9, no. 11 (2006): 1367.

Koyama, T., J. G. McHaffie, P. J. Laurienti, and R. C. Coghill. "The subjective experience of pain: Where expectations become reality." *Proceedings of the National Academy of Science U. S. A*, 102, no. 36 (2005): 12950–55.

For more on Dr. Don Price, who explored the effect of the expectation of pain on people with irritable bowel syndrome, see:

http://www.stoppain.org/for_professionals/compendium/bios/price.asp.

For more on expectations and dopamine, see:

Schultz, W. "The reward signal of midbrain dopamine neurons." *News in Physiological Sciences* 14, no. 6 (1999): 249–55.

———. "Reward signaling by dopamine neurons." *Neuroscientist* 7, no. 4 (2001): 293–302.

Waelti, P., A. Dickinson, and W. Schultz. "Dopamine responses comply with basic assumptions of formal learning theory." *Nature* 412 (2001): 43–48.

For more on the positive effect on mental health of having "rose-colored glasses" on oneself, see:

Taylor, S. E., J. S. Lerner, D. K. Sherman, R. M. Sage, and N. K. McDowell. "Portrait of the self-enhancer: Well adjusted and well liked or maladjusted and friendless?" *Journal of Personality and Social Psychology* 84, no. 1 (2003): 165–76.

Scene 10: Turning Enemies into Friends

For more on the social circuits in the brain, see:

Lieberman, M. D. "Social cognitive neuroscience: A review of core processes." *Annual Review of Psychology* 58 (2007): 259–89.

For more on newborn babies orienting toward a picture of a face above any other picture, see:

Goren, C. C., M. Sarty, and P.Y.K. Wu. "Visual following and pattern discrimination of face-like stimuli by newborn infants." *Pediatrics* 56, no. 4 (1975): 544–49.

For more on child development, see:

Wingert, P., and M. Brant. "Reading Your Baby's Mind." *Newsweek*, August 15, 2005, p. 35.

Jaremka, L. M., S. Gabriel, and M. Carvallo. "What makes us feel the best also makes us feel the worst: The emotional impact of independent and interdependent experiences." *Self and Identity* 10, no.1 2011: 44–63.

For more on how people classify others as friend or foe, right from infancy, see:

Porges, S. W. "Neuroception: A subconscious system for detecting threats and safety." *Zero to Three* 24, no. 5 (2004): 19–24.

For more on relatedness as a primary reward or threat, see:

Baumeister, R. F., and M. R. Leary. "The need to belong: Desire for interpersonal attachments as a fundamental human motivation." *Psychological Bulletin* 117 (1995): 497–529.

Cacioppo, J. T., and B. Patrick. *Loneliness: Human Nature and the Need for Social Connection.* New York: W. W. Norton and Company, 2008.

Carter, E. J., and K. A. Pelphrey. "Friend or foe? Brain systems involved in the perception of dynamic signals of menacing and friendly social approaches." *Journal Social Neuroscience* 3, no. 2 (2008): 151–63.

For more on Maslow's hierarchy of needs, see:

Maslow, A. H. "A theory of human motivation." *Psychological Review* 50 (1943): 370–96.

For more on mirror neurons and empathy, see:

Keysers C., and V. Gazzola. "Towards a unifying neural theory of social cognition." *Progress in Brain Research* 156 (2006): 379–401.

Uddin, L. Q., M. Iacoboni, C. Lange, and J. P. Keenan. "The self and social cognition: The role of cortical midline structures and mirror neurons." *Trends in Cognitive Sciences* 11, no. 4 (2007): 153–57.

For more on how mirror neurons relate to grasping other people's intentions directly, see:

Iacoboni, M., I. Molnar-Szakacs, V. Gallese, G. Buccino, J. C. Mazziotta, and G. Rizzolatti. "Grasping the intentions of others with one's own mirror neuron system." *PloS Biology* 3, no. 3 (2005): 79.

More on how mirror neurons may be involved in autism, see:

Iacoboni, M., and M. Dapretto. "The mirror neuron system and the consequences of its dysfunction." *Nature Reviews Neuroscience* 7 (2006): 924–51.

For more on the ways emotions can ripple out across a group, also called emotional contagion, see:

Barsade, S. G. "The ripple effect: Emotional contagion and its influence on group behavior." *Administrative Science Quarterly* 47 (2002): 644–75.

Wild, B., M. Erb, and M. Bartels. "Are emotions contagious? Evoked emotions while viewing emotionally expressive faces: quality, quantity, time course, and gender differences." *Psychiatry Res.* 102 (2001): 109–24.

For more on how you use one set of brain circuits for thinking about people who you believe are like you and different circuits for others, see:

Mitchell, J. P., C. N. Macrae, and M. R. Banaji. "Dissociable medial prefrontal contributions to judgments of similar and dissimilar others." *Neuron* 50 (2006): 655–63.

For more on how oxytocin increases trust and decreases a natural sense of threat, see:

Kosfeld, M., M. Heinrichs, P. J. Zak, U. Fischbacher, and E. Fehr. "Oxytocin increases trust in humans." *Nature* 435 (2005): 673–76.

Meyer-Lindenberg, A., G. Domes, P. Kirsch, and M. Heinrichs. "Oxytocin and vasopressin in the human brain: Social neuropeptides for translational medicine." *National Reviews Neuroscience* 12, no. 9 (2011): 524–538.

For more on the complexities of oxytocin, see:

De Dreu, C. K., L. L. Greer, M. J. Handgraaf, S. Shalvi, G.A. Van Kleef, M. Baas, et al. "The neuropeptide oxytocin regulates parochial altruism in intergroup conflict among humans." *Science* 328, no. 5984 (2010): 1408–1411.

For more on Daniel Kahneman and social situations being the most rewarding, see:

Kahneman, D. "Objective happiness." In D. Kahneman, E. Deiner, and N. Schwarz, eds., *Well-being: Foundations of Hedonic Psychology*, New York: Russell Sage Foundation, 1999, pp. 3–14.

For more on laughter and oxytocin, see Dr. Robert Provine's book:

Laughter: A Scientific Investigation, New York: Penguin Paperback, 2001.

For more on the brain's innate need for relatedness, see:

Cacioppo, J. T., and B. Patrick. *Loneliness: Human Nature and the Need for Social Connection.* New York: W. W. Norton and Company, 2008.

More on how relatedness decreases stress, see:

Eisenberger, N. I., and M. D. Lieberman. "Why rejection hurts: A common neural alarm system for physical and social pain." *Trends in Cognitive Sciences* 8 (2004): 294–300.

Eisenberger, N. I., J. J. Jarcho, M. D. Lieberman, and B. D. Naliboff. "An experimental study of shared sensitivity to physical pain and social rejection." *Pain* 126 (2006): 132–38.

For more on how memory is affected by speaking out loud, see:

Davachi, L., A. Maril, and A. D. Wagner. "When keeping in mind supports later bringing to mind: Neural markers of phonological rehearsal predict subsequent remembering." *Journal of Cognitive Neuroscience* 13, no. 8 (2001): 1059–70.

For more on how competition reduces empathy, see:

Baumeister, R. F., J. M. Twenge, and C. K. Nuss. "Effects of social exclusion on cognitive processes: Anticipated aloneness reduces intelligent thought." *Journal of Personality and Social Psychology* 83, no. 4 (2002): 817–27.

Shirom, A., S. Toker, Y. Alkaly, O. Jacobson, and R. Balicer. "Work-based predictors of mortality: A 20-year followup of healthy employees." *Health Psychology* 30, no. 3 (2011): 268–275.

Holt-Lunstad, J., T. B. Smith, and J. B. Layton. "Social relationships and mortality risk: A meta-analytic review." *PLoS Medicine* 7, no. 7 (2010). Published online, doi.org/10.1371/journal.pmed.1000316.

Walton, G. M., G. L. Cohen, D. Cwir, and S. J. Spencer. "Mere belonging: The power of social connections." *Journal of Personality and Social Psychology* 102, no. 3 (2012): 513–532.

Dhont, K., A. Roets, and A. Van Hiel, "Opening closed minds: The combined effects of intergroup contact and need for closure on prejudice." *Personality and Social Psychology Bulletin* 37, no. 4 (2011): 514–528.

de Quervain, D. J., U. Fischbacher, V. Treyer, M. Schellhammer, U. Schnyder, A. Buck, and E. Fehr. "The neural basis of altruistic punishment." *Science* 305 (2004): 1254–58.

For more on the release of oxytocin, see:

Kosfeld, M., M. Heinrichs, P. J. Zak, U. Fischbacher, and E. Fehr. "Oxytocin increases trust in humans." *Nature* 435 (2005): 673–76.

For more on Gallup's research, see their Web site, www.gallup.com.

For more on shared goals creating in-groups, see:

Xiao, Y. J. and J. J. Van Bavel. "Sudden shifts in social identity swiftly shape implicit evaluations." *Journal of Experimental Social Psychology* 83, (2019): 55–69.

Scene 11: When Everything Seems Unfair

For more on fairness as a primary reward or threat, see:

Tabibnia, G., and M. D. Lieberman. "Fairness and cooperation are rewarding: Evidence from social cognitive neuroscience." *Annals of the New York Academy of Sciences* 1118 (2007): 90–101.

For more on the ultimatum game, see:

Sanfey, A. G., J. K. Rilling, J. A. Aronson, L. E. Nystrom, and J. D. Cohen. "The neural basis of economic decision-making in the Ultimatum Game." *Science* 300 (2003): 1755–58.

Ideas about the evolutionary basis of fairness come from Steven Pinker's book *How the Mind Works*, New York: W. W. Norton and Company, 1997.

For more on how the teenage brain is less effective than a preteen brain in some activities, see:

Blakemore, S. J. "The social brain of a teenager." *The Psychologist* 20 (2007): 600–602.

McGivern, R. F., J. Andersen, D. Byrd, K. L. Mutter, and J. Reilly. "Cognitive efficiency on a match to sample task decreases at the onset of puberty in children." *Brain and Cognition* 50, no. 1 (2002): 73–89.

For links to more resources on serotonin and fairness, see:

Crockett, M. J., L. Clark, G. Tabibnia, M. D. Lieberman, and T. W. Robbins. "Serotonin modulates behavioral reactions to unfairness." *Science* 320, no. 5884 (2008): 173.

For more on how trust and cooperation increase when people experience fair offers, see:

Decety, J., P. L. Jackson, J. A. Sommerville, T. Chaminade, and A. N. Meltzoff. "The neural bases of cooperation and competition: An fMRI investigation." *Neuroimage* 23 (2004): 744–51.

Rilling, J. K., D. A. Gutman, T. R. Zeh, G. Pagnoni, G. S. Berns, and C. D. Kilts. "A neural basis of social cooperation." *Neuron* 35 (2002): 395–405.

For more on trust and oxytocin, see:

Kosfeld, M., Heinrichs, M., Zak, P. J., Fischbacher, U., and Fehr, E. "Oxytocin increases trust in humans." *Nature* 435 (2005): 673–76.

For more on expression as part of punishment behavior, see:

Xiao, E., and D. Houser. "Emotion expression in human punishment behavior." *Proceedings of the National Academy of Sciences of the United States* 102, no. 20 (2005): 7398–401.

For more on how perceived fairness reduces the difficulties of a downsizing, see:

Brockner, J. "Managing the effects of layoffs on others." *California Management Review* (Winter 1992): 9–27.

Hamel, G., and C. K. Prahalad. "Competing for the future," *Harvard Business Review* (July–August 1994): 122–28.

For more on fairness and organizations, see:

Robbins, J. M., M.T. Ford, and L. E. Tetrick. "Perceived unfairness and employee health: A meta-analytic integration." *Journal of Applied Psychology* 97, no. 2 (2012): 235–272.

For more on how we accept unfairness, see:

Tabibnia, G., A. B. Satpute, and M. D. Lieberman. "The sunny side of fairness: Preference for fairness activates reward circuitry (and disregarding unfairness activates self-control circuitry." *Psychological Science* 19, no. 4 (2008): 339–47.

For more on how we don't experience empathy with people who have been unfair, see:

Seymour, B., T. Singer, and R. Dolan. "The neurobiology of punishment." *Nature Reviews Neuroscience* 8 (2007): 300–311.

Singer, T., B. Seymour, J. P. O'Doherty, K. E. Stephan, R. J. Dolan, and C. D. Frith. "Empathic neural responses are modulated by the perceived fairness of others." *Nature* 439 (2006): 466–69.

For more on how giving to others activates a strong reward response, see:

Moll, J., F. Krueger, R. Zahn, M. Pardini, R. Oliveira-Souza, and J. Grafman. "Human fronto-mesolimbic networks guide decisions about charitable donation." *Proceedings of the National Academy of Science* 103 (2006): 15623–28.

Moll, J., R. Oliveira-Souza, and R. Zahn. "The Neural Basis of Moral Cognition." *Annals of the New York Academy of Sciences* 1124 (2008): 161–80.

Scene 12: The Battle for Status

For more on the study by Chen on the longevity of social pain in comparison to physical pain, see:

Chen, Z., K. D. Williams, J. Fitness, and N. C. Newton. "When hurt will not heal: Exploring the capacity to relive social and physical pain." *Psychological Science* 19, no. 8 (2008): 789–95.

For more on how we maintain specific maps that define status relationships with people, see:

Chiao, J. Y., A. R. Bordeaux, and N. Ambady. "Mental representations of social status." *Cognition* 93, no. 2 (2003): B49–57.

Zink, C., Y. Tong, Q. Chen, D. Bassett, J. Stein, and A. Meyer-Lindenberg. "Know your place: Neural processing of social hierarchy in humans." *Neuron* 58 (2008): 273–83.

Srivastava, S. and C. Anderson. (2011). "Accurate when it counts: Perceiving power and status in social groups." In J. L. Smith, W. Ickes, J. Hall, S. D. Hodges, and W. Gardner, eds. *Managing Interpersonal Sensitivity: Knowing When—and When Not to—Understand Others*, Psychology of Emotions, Motivations, and Actions, New York: Nova Science Publishers, 2011: 41–58.

For more on the impact of threats to status, see:

Eisenberger, N., M. Lieberman, and K. Williams. "Does rejection hurt? An fMRI study of social exclusion." *Science* 302, no. 5643 (2003): 290–92.

Eisenberger, N., and M. Lieberman. "Why rejection hurts: A common neural alarm system for physical and social pain." *Trends in Cognitive Sciences* 8, no. 7 (2004): 294–300.

Eisenberger, N. "The pain of social disconnection: examining the shared neural underpinnings of physical and social pain." *Nature Reviews Neuroscience* 13, no. 6 (2012): 421–434.

Lieberman M., and N. Eisenberg. "The pains and pleasures of social life." *NeuroLeadership Journal* 1 (2008): 38–43.

For more on status in animal communities, see:

Sapolsky, R. *Why Zebra's Don't Get Ulcers*. 3rd ed. New York: Henry Holt and Company, 2004.

A book that illustrates the importance of status is *The Status Syndrome: How Social Standing Affects Our Health and Longevity*, by Michael Marmot, New York: Henry Holt and Company, 2005.

For more on the rewards of status, see:

Izuma, K., D. Saito, and N. Sadato. "Processing of social and monetary rewards in the human striatum." *Neuron* 58, no. 2 (2008): 284–94.

Izuma, K. "The social neuroscience of reputation." *Journal of Neuroscience Research* 72, no. 4 (2012): 283–288.

The study about a computer giving kids positive feedback is:

Scott, Dapretto et al. "Social, Cognitive and Affective Neuroscience." (under review, *Social Cognitive and Affective Neuroscience Journal*, 2008).

For information on the effects on the brain of chronic stress and low socioeconomic status, see:

Evans, G. W., and M. A. Schamberg. "Childhood poverty, chronic stress, and adult working memory." *Proceedings of the National Academy of Sciences of the United States*. Published online, www.pnas.org, March 30, 2009.

The idea for "status hope" being behind the phenomenon of ordinary people doing extraordinary things came from an insight I had while writing.

For more on the link between status and dopamine, see:

Grant, K. A., C. A. Shively, M. A. Nader, R. L. Ehrenkaufer, S. W. Line, T. E. Morton, H. D. Gage, and R. H. Mach. "Effect of social status on striatal dopamine D2 receptor binding characteristics in cynomolgus monkeys assessed with positron emission tomography." *Synapse* 29, no. 1 (1998): 80–83.

For more on the link between testosterone and status, see:

Newman, M. L., J. G. Sellers, and R. A. Josephs. "Testosterone, cognition, and social status." *Hormones and Behavior* 47 (2005): 205–11.

The idea of status being a driver for playing against yourself came from reading about how the brain uses the same circuits to know yourself as you know other people. It was an insight I had that status might explain why setting and achieving your own personal goals might be so motivating.

For more on the study of schadenfreude, see:

Takahashi, H., M. Kato, M. Matsuura, D. Mobbs, T. Suhara, and Y. Okubo. "When your gain is my pain and your pain is my gain: Neural correlates of envy and schadenfreude." *Science* 323, no. 5916 (2009): 937–39.

For more on the SCARF model, see my article in the *NeuroLeadership Journal*:

Rock, D. "SCARF: A brain-based model for collaborating with and influencing others." *NeuroLeadership Journal* 1 (2008): 44–52.

Scene 13: When Other People Lose the Plot

For more on complex dynamic systems and organizations, see Margaret Wheatley's book *Leadership and the New Science: Discovering Order in a Chaotic World*, 3rd ed., San Francisco: Berret-Koehler Publishers, 2006

The idea that we focus on the deficit and problem model more easily because of the desire to avoid uncertainty was one I had while writing the book. It's not been tested, as far as I can tell. The idea came from linking three sets of information: the commonsense notion that we know the past, but the future is uncertain; studies showing that even small uncertainties create a threat response; and other studies showing an automatic avoidance of threatening situations.

For more on how the goals you set determine your perceptions, see:

Ferguson, M. J., and J. A. Bargh. "Liking is for doing: The effects of goal pursuit on automatic evaluation." *Journal of Personality and Social Psychology* 87, no. 5 (2004): 557–72.

For more on priming, see:

Jacoby, L. L. "Perceptual enhancement: Persistent effects of an experience." *Journal of Experimental Psychology: Learning, Memory, and Cognition* 9, no. 1 (1983): 21–38.

For more on how your brain needs to settle on one behavioral approach to every situation, see:

Desimone, R., and J. Duncan. "Neural mechanisms of selective visual attention." *Annual Review of Neuroscience* 18 (1995): 193–222.

For more on how positive affect helps facilitate insight, see:

Subramaniam, K., J. Kounios, T. B. Parrish, and M. Jung-Beeman. "A brain mechanism for facilitation of insight by positive affect." *Journal of Cognitive Neuroscience* 21 (2009): 415–32.

For more on the impact of the "aha" and the "a-duh" experiences, see:

Dougal, S., and J. W. Schooler. "Discovery misattribution: When solving is confused with remembering." *Journal of Experimental Psychology* 136, no. 4 (2007): 577–92.

For more on the impact of insight, see:

Gick, M. L., and R. S. Lockhart. "Cognitive and affective components of insight." In R. J. Sternberg and J. E. Davidson, eds., *The Nature of Insight*, Cambridge, Mass.: MIT Press, 1995, pp. 197–228.

Knoblich, G., S. Ohlsson, and G. Raney. "Resolving impasses in problem solving: An eye movement study." In M. Hahn and S. C. Stoness, eds. *Proceedings of the Twenty-First Annual Conference of the Cognitive Sciences*, Vancouver: Simon Fraser University Press, 1999, pp. 276–81.

Ohlsson has explored the question of what you can do when someone is experiencing an impasse:

Schooler, J. W., and J. Melcher. "The ineffability of insight." In S. M. Smith, T. B. Ward, and R. A. Finke, eds., *The creative cognition approach*, Cambridge Mass.: MIT Press, 1997, pp. 97–133.

Matt Lieberman's process for getting students to give themselves feedback was explained to me in an interview in 2008 at his office at UCLA.

Scene 14: The Culture That Needs to Transform

For more on how changing one's own behavior is hard, Alan Deutschman's book *Change or Die* reveals a figure stated at a conference (2004) on the health care crisis: only one in nine people who underwent heart surgery could change his lifestyle. See:

Deutschman, A. *Change or Die: The Three Keys to Change at Work and in Life*. New York: Collins, 2007.

For a larger argument on the trouble with the carrot-and-stick approach in the workplace, see:

Rock, D., and J. M. Schwartz. "The neuroscience of leadership." *Strategy + Business* 43, 2006. Retrieved from http://www.strategy-business.com/media /file/sb43_06207.pdf.

For more on neural synchrony, see:

Slagter, H. A., A. Lutz, L. L. Greischar, A. D. Francis, S. Nieuwenhuis, and J. M. Davis, et al. "Mental training affects distribution of limited brain resources." *Public Library of Sciences Biology* 5, no. 6 (2007): 138.

For more on how neural synchrony plays an important role in the integration of functional modules in the brain, see:

Ward, L. M., S. M. Doesburg, K. Kitajo, S. E. MacLean, and A. B. Roggeveen. "Neural synchrony in stochastic resonance, attention, and consciousness." *Canadian Journal of Experimental Psychology* 60, no. 4 (2006): 319–26.

For an introduction to solutions-focused therapy, see:

"Solutions-focused brief counselling: An overview." In K. Hunt and M. Robson, eds. *Counselling and Metamorphosis.* Durham, UK: Centre for Studies in Counselling, University of Durham, 1998, pp. 99–106.

For information on appreciative inquiry, see:

Cooperrider, D., and D. Whitney. *Appreciative Inquiry: The Handbook.* Ohio: Lakeshore Publishers, 2002.

For more on Desimone's work on attention, see:

Desimone, R., and J. Duncan. "Neural mechanisms of selective visual attention." *Annual Review of Neuroscience* 18 (1995): 193–222.

For more on gamma band electrical waves and cognition, see:

Kaiser, J., and W. Lutzenberger. "Human gamma-band activity: A window to cognitive processing." *Neuroreport* 16 (2005b): 207–11.

Keil, A., M. M. Müller, W. J. Ray, T. Gruber, and T. Elbert. "Human gamma band activity and perception of a gestalt." *Journal of Neuroscience* 19 (1999): 7152–61.

For more on Hebb's Law, see:

Hebb, D. O. *The Organization of Behavior.* New York: Wiley, 1949.

For many case studies in neuroplasticity, see Normon Doidge's book *The Brain That Changes Itself.* New York: Viking Adult, 2007. Also Jeffrey Schwartz' book *The Mind and the Brain,* New York: Harper Perennial, 2003.

The term *self-directed neuroplasticity* appears in:

Schwartz, J. M., E. Z. Gulliford, J. Stier, and M. Thienemann. "Mindful awareness and self-directed neuroplasticity: Integrating psychospiritual and biological approaches to mental health with a focus on obsessive compulsive disorder." In S. G. Mijares and G. S. Khalsa, eds. *The Psychospiritual Clinician's Handbook: Alternative Methods for Understanding and Treating Mental Disorders.* Binghamton, N.Y.: Haworth Reference Press, 2005, p. 5.

The term *attention density* appears in:

Schwartz, J. M., H. P. Stapp, and M. Beauregard. "Quantum physics in neuroscience and psychology: A neurophysical model of mind–brain interaction." *Philosophical Transactions of the Royal Society,* 2005. Published online, doi:10.1098/rsub200401598, 2005; http:rstb.royalsocietypublishing.org/content/360/1458/1309.abstract.

For more on assimilation of goals, see:

Berkman, E., and M. D. Lieberman. "The neuroscience of goal pursuit: Bridging gaps between theory and data." In G. Moskowitz and H. Grant, eds. *The Psychology of Goals*. New York: Guilford Press, 2009, pp. 98–126.

For more on how the brain learns through story and metaphor, see:

Perry, B. "How the brain learns best." *Instructor* 11, no. 4 (2000): 34–35.

For more on Jim Barrell's work, and approach versus avoidance goals, see:

Price, D. D., and J. J. Barrell. "Some general laws of human emotion: Interrelationships between intensities of desire, expectation, and emotional feeling." *Journal of Personality* 52, no. 4 (2006): 389–409.

For more on the individuality of everyone's experience, see:

Coghill, R. C., J. G. McHaffie, and Y. Yen. "Neural correlates of inter-individual differences in the subjective experience of pain." *Proceedings of the National Academy of Sciences* 100 (2003): 8538–42.

The metaphor about gardening is something that emerged from my own observations. Similar ideas have emerged from research into how to best learn a musical instrument, showing that repetition is a key factor. For more information about maximizing effectiveness in arts education—including learning an instrument—see the Web site for the Dana Foundation: www.dana.org. This site also makes available links, resources, and research regarding the benefits of musical/arts training to cognitive development and psychological health.

The idea of general intelligence and self-awareness being inversely correlated emerges from a series of papers exploring the role of the medial regions versus the lateral (outer) brain regions, and what happens to people with damaged medial prefrontal regions. See:

Beer, J. S., A. P. Shimamura, and R. T. Knight. "Frontal lobe contributions to executive control of cognitive and social behavior." In M. S. Gazzaniga, ed., *The Cognitive Neurosciences III*, Cambridge, Mass.: MIT Press, 2004, pp. 1091–104.

Fox, M. D., A. Z. Snyder, J. L. Vincent, M. Corbetta, D. C. Van Essen, and M. E. Raichle. "The human brain is intrinsically organized into dynamic, anticorrelated functional networks." *PNAS* 102, no. 27 (July 5, 2005): 9673–78.

Gray J. R., C. F. Chabris, and T. S. Braver. "Neural mechanisms of general fluid intelligence." *Nature Neuroscience* (February 18, 2003).

Schnyer, D. M., L. Nicholls, and M. Verfaellie. "The role of VMPC in metamemorial judgments of content retrievability." *Journal of Cognitive Neuroscience* 17 (2005): 832–46.

GLOSSARY

Problems and Decisions

Actors. A metaphor for the information that comes onto, or that you choose to bring onto, the stage; it's what is in your attention.

Alpha band. A slower frequency, connected to the brain not being very active in a particular region.

ARIA / four faces of insight. A model describing the moments before, during, and after insight occurs in the brain. The acronym stands for Awareness, Reflection, Insight, and Action.

Audience. A metaphor for information held in your brain, such as memories and routines.

Basal ganglia. A large region deep in the brain. The basal ganglia (there's more than one of them) control activities that occur with minimal conscious attention, such as walking or driving, or any habitual behavior.

Bottleneck. What occurs when a decision has not been made that holds up other decisions.

Default network. A network of brain regions roughly in the middle front areas of the brain, including the medial prefrontal cortex. It activates when you are not doing much else, and also when you think about yourself and other people. It's a similar idea to the "narrative" network mentioned in the intermission.

Dopamine. One of the two main neurotransmitters involved in stabilizing circuits in the prefrontal cortex (norepinephrine is the other). Dopamine is connected to feeling interested in something. It is important for learning and is present in larger amounts during *toward* emotions such as curiosity.

Embedding. A metaphor for creating circuits in the basal ganglia that can drive behaviors without thinking, or for long-term memories that stay with you.

Gamma band. The fastest brain frequency; gamma band brain waves occur when electrical activity oscillates around forty times a second across the brain. This frequency relates to being conscious. It activates during moments of recognition or insight and during mindfulness mediation.

Impasse. What occurs when you are unable to solve a problem, or are stuck in a small set of solutions. Current solutions may need to be inhibited before the impasse can be breached.

Inhibition. The process of holding information off your stage, of not paying attention to something.

Insight. What occurs when you resolve an impasse and solve a problem in an unexpected way. Insights bring a release of energy and change the brain.

Map. Similar to a circuit or network. The formation of a large number of neurons into a larger pattern held together by synaptic connections.

Norepinephrine. One of two neurotransmitters important for stabilizing circuits in the prefrontal cortex. Think of this as "brain adrenaline." Norepinephrine is central to feeling alert and paying close attention. This neurotransmitter is common in *away* emotions such as anxiety. Reasonable levels are needed for good thinking, but too much and circuits don't hold together well.

Prefrontal cortex. A section of the outer layer of the brain, behind the forehead, involved in many types of executive functioning, planning and coordinating the rest of the brain.

Queue. What occurs behind a bottleneck; decisions can get caught in queues.

Short-term memory. Memory in which information comes into your awareness briefly but doesn't stay for long.

Stage. A metaphor for working memory. (I used this metaphor because it's a way of thinking about working memory using less effort.)

Ventrolateral prefrontal cortex. A region of the prefrontal cortex, beneath the right and left temples, that is important for all types of braking functions, including stopping physical movement and inhibiting emotions or thoughts.

Working memory. Memory that allows you to hold the contents of awareness at any moment. The prefrontal cortex is central to healthy working memory functioning. Working memory is energy-hungry, small and easily overwhelmed.

Intermission

Direct-experience circuit. The circuit that activates when your attention focuses directly on incoming data, such as from external or internal senses.

Director. The term used in this book for mindfulness.

MAAS scale. One of the main tests for everyday mindfulness now used by neuroscientists. Developed by Kirk Brown. Stands for Mindful Attention Awareness Scale.

Mindfulness. The opposite of mindlessness. Involves paying attention, in the present, in an open and accepting way, to whatever experience you are having.

Narrative circuit. The network activated when your attention goes to planning, goal setting, and thinking about the future or past, yourself, or other people. This is similar to the default network discussed in the book.

Social, cognitive, and affective neuroscience. A branch of neuroscience exploring how we interact socially rather than studying individual brains in isolation.

Stay Cool Under Pressure

Allostatic load. A range of markers of stress, including cortisol and adrenaline levels in the blood, as well as immune system activity and blood pressure.

Amygdale. A small brain region that is part of the limbic system, which activates based on the strength of an emotional or motivational response.

Anterior cingulate cortex. A part of the brain that has many functions, including detecting errors within the brain itself, and switching attention.

Away state. An overarching organizing principle to minimize danger and maximize reward. The danger state, called here the "away" state (sometimes called the "avoid" state), involves emotions such as uncertainty, anxiety, and fear. It is easier to activate and more intense as an experience than the toward state. It is a useful state for physical activity, but can reduce prefrontal cortex activation when it increases in intensity.

Cortisol. A hormone used to measure stress levels in the body. Cortisol activates bodily functions to assist with survival, including clotting the blood and reducing digestion. Cortisol levels increase as the away state increases in intensity.

Hippocampus. A brain region central to memory functions, especially longer-term declarative (recallable) memory.

Labeling. The process of putting symbolic words on emotional states. This dampens the activity of the limbic system, while raising prefrontal cortex activation.

Limbic system. A region in the center of the brain important for experiencing emotions, memories, and motivations; includes the amygdale, insula, hippocampus, and orbital frontal cortex.

Reappraisal. The process of changing your interpretation of an event, which also dampens activity of the limbic system.

Suppression. A common approach to managing emotions, which involves attempting to not feel or not show your feelings to others. Tends to backfire, affect memory, and make others uncomfortable.

Toward state. The state of being curious, open, interested in something. It is important for learning, insight, creativity, and change. It is usually less intense and subtler than the away state. Creating a toward state can displace an away state.

Collaborate with Others

Autonomy. Having control or choices. A sense of increasing autonomy is a pleasant reward. A sense of no autonomy can make small stresses overwhelming. Finding choice in a situation increases a perception of autonomy.

Certainty. Your ability to predict the future. Increasing uncertainty is a threat; increasing certainty is a reward (with a few minor exceptions in both cases).

Fairness. The state of being in which people act ethically and appropriately with one another.

Mirror neurons. Neurons in the brain that help us directly experience other people's intentions, motivations, and emotions, by feeling the same way ourselves.

Relatedness. Being safely connected to people around you. It involves sensing if people are friend or foe. Other people are generally foe until proven otherwise.

SCARF model. A model summarizing five social domains that drive human behavior. Each domain can be either a reward or threat at any time. The model comprises Status, Certainty, Autonomy, Relatedness, and Fairness.

Status. Where you are in the social order of communities you are involved in. It is like taking self-esteem and making it relative to other people. A status increase is a reward; a status decrease is a strong threat.

Facilitate Change

Attention density. A way of thinking about and measuring the quality and quantity of attention paid to any particular brain circuit.

Neural synchrony. The way many parts of the brain form a larger circuit and fire in a similar way when close attention is paid to something.

Neuroplasticity. The study of change in the brain, both moment to moment and in the long term.

Problem focus. The automatic way people try to find solutions; sometimes called the deficit model. A focus on problems seems easier because it is more certain, which means it is less of a threat. The approach works well with linear, physical systems, but it breaks down with complex systems such as people and organizations.

Self-directed neuroplasticity. The idea that real change in the brain tends to occur when people rewire their own brains.

INDEX

Page numbers in *italics* refer to illustrations.

ABOUT THE AUTHOR

Dr. David Rock is the cofounder and CEO of the NeuroLeadership Institute (NLI), a global human capital consulting firm. NLI's mission is to make organizations more human through science. At date of press, NLI is working with over forty of the top one hundred global firms, including IBM and Microsoft, and many federal agencies around the world. Over 4,500 companies closely follow and utilize NLI's research in some way, and around one million people managers a year learn models and tools David developed.

In 2006, David coined the term *NeuroLeadership* and became passionate about using neuroscience research to improve individual and organizational performance. His work has since been featured in the *New York Times*, *Bloomberg Businessweek*, the *Globe and Mail*, the *Guardian*, the *Sydney Morning Herald*, and in leadership and human resources publications the world over.

In 2007, David founded the NeuroLeadership Summit, a global initiative bringing neuroscientists and business leaders together. The summit now has over 20,000 people attending virtually every year. He has a professional doctorate in the neuroscience leadership and has published many academic papers on the NeuroLeadership field.

David lives in downtown New York City. In his spare time he likes to surf, snowboard, play the bongos, and laugh at the quirks of his own brain.

This is his fourth book, and others being *Coaching with the Brain in Mind* (Wiley, 2009), *Quiet Leadership* (Collins, 2006), and *Personal Best* (Simon and Schuster Australia, 2003).

For more resources and background on this book, see www.your-brain-at-work.com.

For more on the Neuroscience of Leadership, see www.neuroleadership.com.

For more on David's work personally, see his blog at www.davidrock.net.